T0155335

NEUROMETHODS

Series Editor
Wolfgang Walz
University of Saskatchewan
Saskatoon, SK, Canada

For further volumes:
http://www.springer.com/series/7657

Stem Cell Technologies in Neuroscience

Edited by

Amit K. Srivastava

*Department of Pediatric Surgery, Children's Program in Regenerative Medicine,
University of Texas Health Science Center at Houston, McGovern Medical School, Houston, Texas, USA*

Evan Y. Snyder

*Center for Stem Cells and Regenerative Medicine,
Sanford Burnham Prebys Medical Discovery Institute, La Jolla, CA, USA*

Yang D. Teng

*Departments of PM&R and Neurosurgery, Harvard Medical School,
Spaulding Rehabilitation Network and Brigham & Women's Hospital, Boston, Massachusetts, USA*

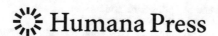 Humana Press

Editors
Amit K. Srivastava
Department of Pediatric Surgery
Children's Program in Regenerative
    Medicine
University of Texas Health
    Science Center at Houston
McGovern Medical School
Houston, TX, USA

Evan Y. Snyder
Center for Stem Cells and Regenerative Medicine
Sanford Burnham Prebys Medical Discovery
    Institute
La Jolla, CA, USA

Yang D. Teng
Departments of PM&R and Neurosurgery
Harvard Medical School
Spaulding Rehabilitation Network and
    Brigham & Women's Hospital
Boston, MA, USA

ISSN 0893-2336   ISSN 1940-6045   (electronic)
Neuromethods
ISBN 978-1-4939-8371-1   ISBN 978-1-4939-7024-7   (eBook)
DOI 10.1007/978-1-4939-7024-7

Cover illustration: Green fluorescent protein expressing human inducible pluripotent stem cells (hiPSCs: green) genetically engineered to produce the herpes simplex virus thymidine kinase (HSV-TK), formed extensive gap junctions (connexin 43 immunoreactivity: red dots) with Nanog positive (purple) U87 human (Grade IV) glioblastoma multiforme cells. (Original confocal z-stack imaging data of Teng Laboratory, Department of PM&R, Harvard Medical School. Z-axis scale: 0.50 μm. *All rights reserved*).

Printed on acid-free paper

This Humana Press imprint is published by Springer Nature
The registered company is Springer Science+Business Media LLC
The registered company address is: 233 Spring Street, New York, NY 10013, U.S.A.

Preface to the Series

Experimental life sciences have two basic foundations: concepts and tools. The *Neuromethods* series focuses on the tools and techniques unique to the investigation of the nervous system and excitable cells. It will not, however, shortchange the concept side of things as care has been taken to integrate these tools within the context of the concepts and questions under investigation. In this way, the series is unique in that it not only collects protocols but also includes theoretical background information and critiques which led to the methods and their development. Thus, it gives the reader a better understanding of the origin of the techniques and their potential future development. The *Neuromethods* publishing program strikes a balance between recent and exciting developments like those concerning new animal models of disease, imaging, in vivo methods, and more established techniques, including immunocytochemistry and electrophysiological technologies. New trainees in neurosciences still need a sound footing in these older methods in order to apply a critical approach to their results.

Under the guidance of its founders, Alan Boulton and Glen Baker, the *Neuromethods* series has been a success since its first volume published through Humana Press in 1985. The series continues to flourish through many changes over the years. It is now published under the umbrella of Springer Protocols. While methods involving brain research have changed a lot since the series started, the publishing environment and technology have changed even more radically. Neuromethods has the distinct layout and style of the Springer Protocols program, designed specifically for readability and ease of reference in a laboratory setting.

The careful application of methods is potentially the most important step in the process of scientific inquiry. In the past, new methodologies led the way in developing new disciplines in the biological and medical sciences. For example, physiology emerged out of Anatomy in the nineteenth century by harnessing new methods based on the newly discovered phenomenon of electricity. Nowadays, the relationships between disciplines and methods are more complex. Methods are now widely shared between disciplines and research areas. New developments in electronic publishing make it possible for scientists that encounter new methods to quickly find sources of information electronically. The design of individual volumes and chapters in this series takes this new access technology into account. Springer Protocols makes it possible to download single protocols separately. In addition, Springer makes its print-on-demand technology available globally. A print copy can therefore be acquired quickly and for a competitive price anywhere in the world.

Saskatoon, Canada   *Wolfgang Walz*

Preface

In the late 1980s to the early 1990s, the discovery of multipotent engraftable stem cells within tissues that were previously thought to be immutable and rigid (e.g., the central nervous system and heart) gave birth to the field of regenerative medicine. By the late 1990s, the ability to isolate cells that were pluripotent expanded the repertoire of cell types that could be generated and transplanted. The discovery that somatic cells from living patients could be reprogrammed back to pluripotence or, with genetic engineering, even transdifferentiated to another somatic cell type without being reprogrammed back to pluripotence further expanded the "palette" from which practitioners of regenerative medicine could choose appropriate cells. Paralleling the discovery of the range of cell types that might be used translationally was a growing appreciation for the multiple therapeutic actions of those cells. It came to be realized that not only might cells be used for replacement of degenerating or missing cells but also they might be used for molecular therapy as chaperone cells for restoring homeostasis to disordered systems. Such actions might be via diffusible factors, through the intercellular transfer of molecules via gap junctions or by the engulfment of microvesicles, or by changing the differentiation fate of host stem/progenitor cells. Contemporaneously with these discoveries was a growing realization of the multiple pathological processes ongoing for each disease, to which the multiple modal actions of the stem cells might be mapped. Translational stem cell biologists came to realize that they must tailor the choice of a particular stem cell to the often-unique needs of a particular disease state. In recognition of growing excitement and potential of stem cells, there is a need to provide comprehensive information and detailed laboratory protocols used in stem cell biology.

Houston, TX, USA   *Amit K. Srivastava*
La Jolla, CA, USA   *Evan Y. Snyder*
Boston, MA, USA   *Yang D. Teng*

Contents

Contributors

JAMIE E. ANDERSON • *Department of Neurosurgery, Harvard Medical School and Brigham and Women's Hospital, Boston, MA, USA; Division of SCI Research, VA Boston Healthcare System, Boston, MA, USA*

MICHAEL C. BREADMORE • *Australia Centre for Research on Separation Science (ACROSS), School of Physical Sciences, University of Tasmania, Hobart, TAS, Australia*

OLIVER BRÜSTLE • *Institute of Reconstructive Neurobiology, LIFE and Brain Center, University of Bonn Medical Faculty, Bonn, Germany*

LEONORA BUZANSKA • *Stem Cell Bioengineering Unit, Mosakowski Medical Research Centre, Polish Academy of Sciences, Warsaw, Poland*

MALGORZATA BUREK • *Department of Anesthesia and Critical Care, University of Würzburg, Würzburg, Germany*

ENJANA BYLYKBASHI • *Genetics and Aging Research Unit, MassGeneral Institute for Neurodegenerative Disease, Massachusetts General Hospital, Harvard Medical School, Charlestown, MA, USA*

LINZHAO CHENG • *Cellular and Molecular Medicine Program, The Johns Hopkins University School of Medicine, Baltimore, MD, USA; Division of Hematology, Department of Medicine, The Johns Hopkins University School of Medicine, Baltimore, MD, USA; Stem Cell Program, Institute for Cell Engineering, The Johns Hopkins University School of Medicine, Baltimore, MD, USA*

LIQUAN WU • *Department of Physical Medicine and Rehabilitation, Harvard Medical School and Spaulding Rehabilitation Hospital, Boston, MA, USA; Division of SCI Research, VA Boston Healthcare System, Boston, MA, USA*

SE HOON CHOI • *Genetics and Aging Research Unit, MassGeneral Institute for Neurodegenerative Disease, Massachusetts General Hospital, Harvard Medical School, Charlestown, MA, USA*

JULIANA CORRÊA-VELLOSO • *Department of Biochemistry, Institute of Chemistry, University of São Paulo, São Paulo, Brazil*

CARLA D'AVANZO • *Genetics and Aging Research Unit, MassGeneral institute for Neurodegenerative Disease, Massachusetts General Hospital, Harvard Medical School, Charlestown, MA, USA*

RALF DAHM • *Department of Biology, University of Padova, Padova, Italy*

TRACEY C. DICKSON • *Menzies Institute for Medical Research, University of Tasmania, Hobart, TAS, Australia*

ROLF O. EHRHARDT • *MedCision, Inc., San Rafael, CA, USA*

CAROLA Y. FÖRSTER • *Department of Anesthesia and Critical Care, University of Würzburg, Würzburg, Germany*

TALITA GLASER • *Department of Biochemistry, Institute of Chemistry, University of São Paulo, São Paulo, Brazil*

ROSANNE M. GUIJT • *School of Medicine and ACROSS, University of Tasmania, Hobart, TAS, Australia*

INBO HAN • *Departments of PM&R and Neurosurgery, Harvard Medical School and Spaulding Rehabilitation Network, Boston, MA, USA; Division of SCI Research, VA Boston Healthcare System, Boston, MA, USA; Department of Neurosurgery, Cha University, Seoul, South Korea*

DANYANG HE • *Divisions of Experimental Hematology and Cancer Biology & Developmental Biology, Department of Pediatrics, Cincinnati Children's Hospital Medical Center, Cincinnati, OH, USA*

MATTHIAS HEBISCH • *Institute of Reconstructive Neurobiology, LIFE and Brain Center, University of Bonn Medical Faculty, Bonn, Germany; German Center for Neurodegenerative Diseases (DZNE), Bonn, Germany*

MARK E. HESTER • *Center for Perinatal Research, The Research Institute at Nationwide Children's Hospital, Columbus, OH, USA*

ALEXIS B. HOOD • *Center for Perinatal Research, The Research Institute at Nationwide Children's Hospital, Columbus, OH, USA*

MIROSLAW JANOWSKI • *Division of MR Research, Russell H. Morgan Department of Radiology and Radiological Science, Cellular Imaging Section and Vascular Biology Program, Institute for Cell Engineering, The Johns Hopkins University, Baltimore, MD, USA; NeuroRepair Department, Mossakowski Medical Research Centre, Polish Academy of Sciences, Warsaw, Poland*

ROBERT T. KARL • *Department of Genetics and Genome Sciences, Case Western Reserve University School of Medicine, Cleveland, OH, USA*

DOO YEON KIM • *Genetics and Aging Research Unit, MassGeneral Institute for Neurodegenerative Disease, Massachusetts General Hospital, Harvard Medical School, Charlestown, MA, USA*

YOUNG HYE KIM • *Division off Bioconvergence Analysis, Korea Basic Science Institute, Cheongju, Chungbuk, Republic of Korea*

ANNA E. KING • *Wicking Dementia Research and Education Centre, School of Medicine, University of Tasmania, Hobart, TAS, Australia*

ERIC J. KUNKEL • *MedCision, Inc., San Rafael, CA, USA*

ANGELA M. LAGER • *Department of Genetics and Genome Sciences, Case Western Reserve University School of Medicine, Cleveland, OH, USA*

Q. RICHARD LU • *Divisions of Experimental Hematology and Cancer Biology & Developmental Biology, Department of Pediatrics, Cincinnati Children's Hospital Medical Center, Cincinnati, OH, USA*

PAOLO MACCHI • *Laboratory of Molecular and Cellular Neurobiology, Centre for Integrative Biology (CIBIO), University of Trento, Trento, Italy*

VASILIKI MAHAIRAKI • *Department of Neurology and Neurosurgery, The Johns Hopkins University School of Medicine, Baltimore, MD, USA*

BRADLEY MEYER • *Division of Experimental Hematology and Cancer Biology & Developmental Biology, Department of Pediatrics, Cincinnati Children's Hospital Medical Center, Cincinnati, OH, USA*

DILLON C. MUTH • *Department of Molecular and Comparative Pathobiology, The Johns Hopkins University School of Medicine, Baltimore, MD, USA; Cellular and Molecular Medicine Program, The Johns Hopkins University School of Medicine, Baltimore, MD, USA*

FADI J. NAJM • *Department of Genetics and Genome Sciences, Case Western Reserve University School of Medicine, Cleveland, OH, USA*

ÁGATHA OLIVEIRA-GIACOMELLI • *Department of Biochemistry, Institute of Chemistry, University of São Paulo, São Paulo, Brazil*

ANA RUIZ • *European Commission, Joint Research Centre, Institute for Health and Consumer Protection, Ispra, Italy; European Commission, Joint Research Centre, Institute for Energy and Transport, Petten, The Netherlands*

ANNALISA ROSSI • *Laboratory of Molecular and Cellular Neurobiology, Centre for Integrative Biology (CIBIO), University of Trento, Trento, Italy*

FRANÇOIS ROSSI • *European Commission, Joint Research Centre, Institute for Health and Consumer Protection, Ispra, Italy*

ELLAINE SALVADOR • *Department of Anesthesia and Critical Care, University of Würzburg, Würzburg, Germany*

PHILIP H. SCHWARTZ • *National Human Neural Stem Cell Resource, Children's Hospital of Orange County Research Institute, Orange, CA, USA*

EVAN Y. SNYDER • *Center for Stem Cells and Regenerative Medicine, Sanford Burnham Prebys Medical Discovery Institute, La Jolla, CA, USA*

RULDOLPH E. TANZI • *Genetics and Aging Research Unit, MassGeneral Institute for Neurodegenerative Disease, Massachusetts General Hospital, Harvard Medical School, Charlestown, MA, USA*

YANG D. TENG • *Departments of PM&R and Neurosurgery, Harvard Medical School, Spaulding Rehabilitation Network and Brigham & Women's Hospital, Boston, MA, USA*

PAUL J. TESAR • *Department of Genetics and Genome Sciences, Case Western Reserve University School of Medicine, Cleveland, OH, USA*

MARIA L. THOMPSON • *MedCision, Inc., San Rafael, CA, USA*

HENNING ULRICH • *Department of Biochemistry, Institute of Chemistry, University of São Paulo, São Paulo, Brazil*

PIOTR WALCZAK • *Division of MR Research, Rusell H. Morgan Department of Radiological Science, Cellular Imaging Section and Vascular Biology Program, Institute for Cell Engineering, The Johns Hopkins University, Baltimore, MD, USA; Department of Neurology and Neurosurgery, Faculty of Medical Sciences, University of Warmia and Mazury, Olsztyn, Poland*

DAVID E. WEINSTEIN • *Rivertown Therapeutics, Inc., New York, NY, USA*

ROBIN L. WESSELSCHMIDT • *National Human Neural Stem Cell Resource, Children's Hospital of Orange County Research Institute, Orange, CA, USA*

KENNETH W. WITWER • *Department of Molecular and Comparative Pathobiology, The Johns Hopkins University School of Medicine, Baltimore, MD, USA; Cellular and Molecular Medicine Program, The Johns Hopkins University School of Medicine, Baltimore, MD, USA; Department of Neurology, The Johns Hopkins University School of Medicine, Baltimore, MD, USA*

YIING CHIING YAP • *Menzies Institute for Medical Research, University of Tasmania, Hobart, TAS, Australia; Wicking Dementia Research and Education Centre, School of Medicine, University of Tasmania, Hobart, TAS, Australia; Australia Centre for Research on Separation Sciences (ACROSS), School of Physical Sciences, University of Tasmania, Hobart, TAS, Australia; School of Medicine and ACROSS, University of Tasmania, Hobart, TAS, Australia*

XIANG ZENG • *Department of Neurosurgery, Harvard Medical School and Brigham and Women's Hospital, Boston, MA, USA; Division of SCI Research, VA Boston Healthcare System, Boston, MA, USA*

ZEZHOU ZHAO • *Department of Molecular and Comparative Pathobiology, The Johns Hopkins University School of Medicine, Baltimore, MD, USA*

MARZENA ZYCHOWICZ • *Stem Cell Bioengineering Unit, Mosakowski Medical Research Centre, Polish Academy of Sciences, Warsaw, Poland*

Chapter 1

Three-Dimensional Cultures of Human Neural Stem Cells: An Application for Modeling Alzheimer's Disease Pathogenesis

Se Hoon Choi, Carla D'Avanzo, Young Hye Kim, Enjana Bylykbashi, Matthias Hebisch, Oliver Brüstle, Ruldolph E. Tanzi, and Doo Yeon Kim

Abstract

A three-dimensional (3D) cell culture system allows cells to grow and attach to their environments in all three directions. As opposed to the conventional method of growing cells in a flat environment on a petri dish (2D), the 3D cultures can closely mimic the in vivo cellular environment. Recently, we developed a novel cell culture model system of Alzheimer's disease (AD) by employing a 3D culture setting. In this model, human neural progenitor cells harboring multiple familial AD (fAD) mutations were differentiated in a 3D Matrigel. Using this 3D culture model, we have shown, for the first time, that fAD mutations can induce β-amyloid plaque and tau tangle pathologies. These two major pathological hallmarks of AD have not yet been achieved in current AD mouse models. In this chapter, we describe instructions and troubleshooting for establishing 3D human cell culture models that can be used to analyze β-amyloid and tau pathology in a dish.

Key words Alzheimer's disease, Three-dimensional cell culture, Human neural progenitors, ReN cells, lt-NES cells, Extracellular Aβ deposits, Tauopathy

1   Introduction: Alzheimer's Disease

Alzheimer's disease (AD) is the most common neurodegenerative disease characterized by progressive memory loss and overall cognitive decline. The two key pathological hallmarks of AD are β-amyloid plaques and neurofibrillary tangles (NFTs). Amyloid plaques are extracellular deposits of amyloid-β (Aβ) peptides, which are liberated from larger amyloid precursor proteins (APP) through sequential cleavage by β- and γ-secretases. NFTs are filamentous aggregates of hyperphosphorylated tau protein found inside cells [1]. According to the amyloid cascade hypothesis, the accumulation of pathogenic Aβ species is the initial pathological trigger in the disease, which subsequently leads to the formation of NFTs, neuronal cell death and, ultimately, dementia.

Amit K. Srivastava et al. (eds.), *Stem Cell Technologies in Neuroscience*, Neuromethods, vol. 126, DOI 10.1007/978-1-4939-7024-7_1, © Springer Science+Business Media LLC 2017

1.1 Current AD Models and their Limitations

Cellular and animal model systems are essential to investigate pathogenic mechanisms and to test strategies for treatments. For AD, mouse models have largely contributed to the study of pathogenic mechanisms of AD and to test candidate AD drugs before human trials. The majority of AD mouse models harbor single or multiple human familial AD (fAD) mutations and show robust Aβ-amyloid plaques and Aβ-driven deficits, including synaptic and memory deficits. However, fAD mouse models do not exhibit Aβ-driven tau tangle formation, which is predicted by the amyloid cascade hypothesis [2]. Mouse models with both fAD mutations and frontotemporal dementia (FTD) tau mutations were able to recapitulate both Aβ plaques and tau tangles; most of the tau pathologies in these models result from strong FTD mutations, not fAD mutation [3, 4]. Also, animal models have fundamental species-specific differences in genome and protein composition, likely precluding the recapitulation of *bona fide* AD pathological events, which occur in aged human brains. Immortalized cell lines and primary rodent cell cultures, although having been widely used for elucidating the molecular mechanism of AD pathogenesis, do not adequately represent the environmental features of the native brain environment.

Recent stem cell reprogramming technology has made it possible to generate human neuronal cells harboring disease-specific genetic lesions of AD patients [5–12]. These neurons would have the genetic background of human AD patients, and thus, are expected to recapitulate AD pathogenesis precisely. However, similar to immortalized cell lines and primary rodent cells, they are not able to generate robust extracellular β-amyloid aggregation or NFT pathologies, including aggregation or paired helical filament (PHF) formation, in conventional two-dimensional (2D) culture systems.

1.2 3D Neural Stem Cell Culture Models of AD

Lack of robust AD pathologies in the current in vitro models might be due to two factors: (1) the relatively low levels of pathogenic Aβ species, (2) the fundamental limitation of conventional 2D cell culture systems to recapitulate the highly complex and dynamic 3D brain environment. In conventional 2D cell culture systems, the secreted Aβ might diffuse into the relatively large volume of cell culture media, and is likely to be removed during the regular media changes before aggregation [2, 13, 14]. In contrast, 3D neural cell culture architecture would provide a brain tissue-like closed environment, which would seed and accelerate Aβ deposition by limiting the diffusion of secreted Aβ into the cell culture medium, as well as by providing local niches that enable enrichment of Aβ levels high enough for the initiation of aggregation [15–18].

Recently, we developed a 3D culture model of human neural progenitor cells (hNPCs) overexpressing fAD-causing *APP* and

Presenilin 1 (*PS1*) mutations [13, 14]. To establish the system, we chose the ReNcell VM cell line (ReN cells), an immortalized hNPC line that readily differentiates into neurons and glial cells. For fAD mutations, we chose the *APP Swedish* and *London* mutations (K670N/M671L and V717I, respectively), and *PSEN1(ΔE9)* to produce high levels of Aβ40 and 42, and to elevate the Aβ42/Aβ40 ratio. Matrigel was chosen as a 3D hydrogel since it easily solidifies with cells upon moderate thermal change and contains brain extracellular matrix proteins [19]. After differentiating hNPCs with fAD mutations in 3D Matrigel for 6 weeks, we were able to detect robust extracellular Aβ deposits and sodium dodecyl sulfate (SDS)-resistant Aβ aggregates [13, 14]. More importantly, robust accumulations of hyperphosphorylated tau proteins were evident in the somatodendritic compartments of fAD neurons without the need for additional tau mutations (i.e. FTD mutations) [13, 14]. Immunoelectron microscopy also revealed the presence of detergent-insoluble filamentous structures labeled by tau antibodies [13]. All these observations clearly demonstrated robust Aβ plaques and NFT-like pathologies in our 3D-AD models.

Besides the 3D neural cell culture systems being especially suitable for reconstituting extracellular aggregation of Aβ, they are superior for recapitulating in vivo brain environments and can accelerate neuronal differentiation and neural network formation [16–18, 20, 21]. Neurons grown in a 2D environment are able to spread freely across the surface of the glass or plastic surface, but they have no support for vertical growth leading to unintended apical–basal polarity rather than the physiological stellar morphology of neurons [22]. Indeed, we have also shown that neurons in our 3D cultures dramatically elevated maturity marker expression compared to 2D-differentiated cells [13]. Strikingly, we found that neurons in 3D culture condition express elevated levels of 4-repeat adult tau (4R tau) isoforms, which are essential for recapitulating the NFT pathology [13]. High levels of the 4R tau isoform would also contribute to the robust tauopathy observed in our 3D human neural cell model of AD.

In this chapter, we comprehensively describe protocols to establish our human 3D culture system that is directly applicable to recapitulate AD pathogenesis, including Aβ plaques and tau tangles. The current 3D culture systems are optimized for ReN cell lines [13, 14]. In addition, we present and describe a slightly modified 3D culture of physiological human neuronal cells, human induced pluripotent stem cell (iPSC)-derived long-term self-renewing neuroepithelial stem cells (It-NES).

2   Materials

2.1   fAD ReN Cells

The immortalized human neural progenitor cell line, ReNcell VM (ReN cell), was purchased from EMD Millipore (Billerica, MA). Accutase solution was obtained from Life Technologies (Waltham, MA). ReN proliferation medium is DMEM/F12 media supplemented with 2 μg/mL heparin, 2% (v/v) B27 neural supplement, 20 μg/mL EGF, 20 μg/mL bFGF, and 1% (v/v) penicillin/streptomycin/amphotericin-b solution. ReN proliferation medium and Accutase solution were pre-warmed in a water bath (37 °C) for ~10 min. The lentiviral constructs encoding human APP and human PS1 with fAD mutations have been described in detail in our previous publications [13, 14].

2.2   Preparation and Differentiation of ReN Cells in 3D Matrigel

BD Matrigel was purchased from BD Biosciences (San Jose, CA). Matrigel stocks were kept in a −80 °C freezer. A stock was taken out and placed in 4 °C refrigerator 1 day before use and kept on ice until it was plated with the cells (Matrigel tends to solidify above 10 °C). Each lot of Matrigel has a different protein concentration and therefore needs to be pretested. We generally test 1:10, 1:15, and 1:20 dilutions and pick the best dilution rate for each lot. Before transferring Matrigel, pipettes should be chilled by pipetting cold differentiation medium back and forth. ReN cell differentiation medium was DMEM/F12 supplemented with 2 μg/mL heparin, 2% (v/v) B27 neural supplement, and 1% (v/v) penicillin/streptomycin/amphotericin-b solution without growth factors. The culture plates used for 2D and 3D culture of ReN cells were: Lab-Tek II 8-well chambered cover glass slides (Thermo Scientific, Rockford, IL), Greiner Bio-One 96-well uClear black TC plates, flat bottom (Greiner Bio-One, Monroe, NC), Cell culture inserts, 0.4 μm pore size (Fisher Scientific/Falcon, Pittsburg, PA), 24-well Falcon companion plates for cell culture inserts (Fisher Scientific/Falcon, Pittsburg, PA) and 6-well cell culture plates (Fisher Scientific/Corning life sciences, Pittsburg, PA).

2.3   Differential Detergent Extraction and Detection of Aggregated Aβ and Phospho-tau Species

Tris-buffered saline (TBS) extraction buffer contains 50 mM Tris (pH 7.4), 150 mM NaCl, 1 mM NaVO$_3$, 1 mM NaF, a protease inhibitor mixture (Roche Molecular Biochemicals, Indianapolis, IN), a phosphatase inhibitor cocktail (Thermo Scientific, Rockford, IL), 2 mM PNT (EMD Millipore, Billerica, MA) and 1 mM phenylmethylsulfonyl fluoride (PMSF, Sigma-Aldrich, St. Louis, MO). 2% SDS extraction buffer contains 50 mM Tris (pH 7.4), 150 mM NaCl, 2% SDS, 1% TritonX-100, 1 mM NaVO$_3$, 1 mM NaF, a protease inhibitor mixture, a phosphatase inhibitor cocktail, 2 mM PNT and 1 mM PMSF. 1% sarkosyl/RIPA buffer contains 50 mM Tris (pH 7.4), 150 mM NaCl, 0.5% w/v sodium deoxycholate, 2% v/v NP-40, 1% w/v N-Lauroylsarcosine, a protease inhibitor mixture, a phosphatase inhibitor cocktail, 1 mM NaF, 1 mM NaVO$_3$, 2 mM PNT, and 1 mM PMSF.

Primary antibodies were used at the following dilutions: 6E10 anti-amyloid-β (1:300, Covance, Dedham, MA); anti-PS1 (1:1000, Cell Signaling Technology, Danvers, MA); anti-α-tubulin (1:1000, Cell Signaling Technology, Danvers, MA); anti-CNPase (1:1000, Cell Signaling Technology, Danvers, MA); anti-β secretase 1 (1:1000, Cell Signaling Technology, Danvers, MA); C66 APP C-terminal antibody (1:2000, a gift from Dr. Kovacs); AT8 anti-p-tau (1:100, Millipore, Billerica, MA); PHF1 anti-p-tau (1:500, a gift from Dr. Peter Davies); anti-total tau (1:2000, DAKO, Fort Collins, CO); anti-MAP2 (1:500, Millipore, Billerica, MA; 1:200, Cell Signaling Technology, Danvers, MA); anti-NCAM (1:1000, Cell Signaling Technology, Danvers, MA); anti-synapsin I (1:500, Cell Signaling Technology, Danvers, MA); anti-HSP70 (1:1000, Enzo Life Sciences, Farmingdale, NY), and anti-human mitochondrial antigen (1:500, Millipore, Billerica, MA).

2.4 Reagents and Antibodies for Immuno-staining

Blocking solution contains 50 mM Tris (pH 7.4), 0.1% Tween 20, 4% donkey serum, 1% bovine serum albumin (BSA), 0.1% gelatin, and 0.3 M glycine. TBST buffer is TBS buffer containing 0.1% (v/v) Tween 20. The antibodies for fluorescence staining that we used are: 3D6 anti-amyloid-β antibody (1:500, a gift from Lilly); anti-β-tubulin type III (TUJ1, 1:200, Abcam, Cambridge, MA); anti-GFAP antibody (1:2000, DAKO, Fort Collins, CO; 1:200, Antibodies Incorporated, Davis, CA); AT8 anti-p-tau antibody (1:40, Thermo Scientific, Rockford, IL); PHF1 anti-phospho tau antibody (1:1000, a gift from Peter Davies); anti-tau46 antibody (1:200, Cell Signaling Technology, Danvers, MA); anti-GluR2 (1:100, Antibodies Incorporated, Davis, CA); anti-MAP2 antibodies (1:400, Millipore, Billerica, MA; 1:200, Cell Signaling Technology, Danvers, MA); anti-tyrosine hydroxylase (1:100, Cell Signaling Technology, Danvers, MA); anti-NR2B (1:100, Antibodies Incorporated, Davis, CA); anti-GABA(B)R2 (1:100, Cell Signaling Technology, Danvers, MA); Alexa Fluor 350/488/568 anti-mouse, -rabbit, and -chicken secondary antibodies (1:200, Life Technologies, Waltham, MA).

For immunohistochemistry, the following antibodies were used: BA27 anti-Aβ40 antibody horseradish peroxidase (HRP)-conjugate (1:2, Wako Chemicals, Richmond, VA); BC05 anti-Aβ42 antibody HRP-conjugate (1:2, Wako Chemicals, Richmond, VA); AT8 anti-p-tau antibody (1:40, Thermo Scientific, Rockford, IL); PHF1 anti-p-tau antibody (1:1000, courtesy of P. Davies), anti-MAP2 antibody (1:200, Cell Signaling Technology, Danvers, MA); ImmPRESS anti-mouse and -rabbit IgG HRP polymer conjugates (1:2, Vector Laboratories, Burlingame, CA). ImmPRESS Peroxidase Polymer Detection Kits and ImmPACT DAB Peroxidase Substrate kits were purchased from Vector Laboratories (Burlingame, CA).

2.5 iPSC and It-NES Cultures

Plastic cell culture dishes (Sigma-Aldrich, St. Louis, MO) were coated either with Matrigel (1:30 in DMEM/F12) for iPSC culture or sequentially with poly-ornithine and laminin (PO/Ln, Sigma-Aldrich, St. Louis, MO) for It-NES. Uncoated plastic petri dishes (Falcon, Pittsburg, PA) were used for free-floating cultures. IPSCs were cultured in E8 medium [23], containing Rho-kinase inhibitor (Y-27632, Tocris Bioscience, Bristol, United Kingdom) in specific steps. N2 medium consists of DMEM/F12 (Life Technologies, Waltham, MA) with N2 supplement (1:100, GE Healthcare, Wilmington, MA) and glucose (1.6 mg/mL, Sigma-Aldrich, St. Louis, MO). For It-NES cultivation, N2 is supplemented with EGF, FGF (10 ng/mL each, R&D Systems, Minneapolis, MN), and B27 supplement (1:1000, Life Technologies, Waltham, MA). Neural differentiation medium is prepared using 50% N2 medium, 50% Neurobasal medium (Life Technologies, Waltham, MA), and 1:50 B27 supplement. To dislodge cells, we used 0.5 mM EDTA (Sigma-Aldrich, St. Louis, MO) in Ca^{2+}/Mg^{2+}-free PBS (Gibco) [24] for iPSC colonies, Accutase (Life Technologies, Waltham, MA) for single-cell iPSCs, or 0.05% trypsin and trypsin inhibitor (both from Life Technologies, Waltham, MA) for lt-NES cells.

3 Methods

3.1 Generation of ReN Cells Expressing High Levels of Aβ: Cell Maintenance, Viral Infection, and Enrichment of High-Expressing Transgenic ReN Cells

ReN cells were plated onto Matrigel-coated T25 cell culture flasks and maintained in ReN cell proliferation media in a 37 °C CO_2 cell culture incubator. Medium replacement was performed every 3 days until the cells reached ~95% confluency. To passage, the confluent cells were first washed with ~3 mL of D-PBS, which was then carefully aspirated. To detach the cells, 0.5 mL of Accutase were added and after a 3–5 min incubation in a 37 °C CO_2 incubator, the flask was firmly tapped in order to dislodge still-attached cells. After adding 3 mL of pre-warmed ReN proliferation medium, the cell suspension was transferred to a 15 mL sterile conical tube and spun down at $2000 \times g$ for 3 min. The supernatant was then removed and the cell pellet was resuspended in 12 mL of proliferation medium. 4 mL of this ReN cell suspension was dispensed into each Matrigel-coated T25 flask, which was then incubated in a 37 °C CO_2 incubator until confluent again.

The construction of fAD lentiviral constructs encoding human APP and human PS1 with fAD mutations has been described in detail in our publications [13, 14]. In brief, APP with *Swedish/London* FAD mutations (APPSL) and/or PS1 with delta E9 (PS1ΔE9) cDNAs were PCR-amplified with Pfu (New England Biolabs, Ipswich, MA) and cloned into lentiviral polycistronic CSCW-GFP or CSCW-mCherry vectors (Massachusetts General Hospital (MGH) viral core) to generate CSCW-APPSL-GFP and

CSCW-PS1ΔE9-mCherry. To make the CSCW-APP-PS1ΔE9-mCherry vector, the APPSL-IRES segment of CSCW-APPSL-GFP vector were PCR-amplified, digested with NheI and cloned into NheI site of CSCW-PS1ΔE9-mCherry vector.

To generate fAD ReN cells, the proliferating ReN cells were transfected with CSCW-APPSL-GFP, CSCW-PS1ΔE9-mCherry or CSCW-APPSL-PS1ΔE9-mCherry either alone or together. ReN cells with only CSCW-GFP or -mCherry served as controls. To prepare ReN cells for infection, the cells from confluent T25 flasks were extricated as described above. The cell pellet was resuspended in 12 mL of pre-warmed proliferation medium, of which 2 mL were dispensed into each well of a Matrigel-coated 6-well plate and allowed to settle overnight. Upon reaching 70–80% confluency, 6×10^6 TU (Transducing Units) of viral particles were added per well to achieve an approximate M.O.I. (Multiplicity of Infection) of 1. After gentle mixing, the plate was incubated overnight. Expression of the transgenes can be expected 48 h after infection and should be detectable by fluorescence microscopy. To increase the expression of pathogenic Aβ species, we generated fAD ReN cells expressing high levels of both APPSL and PS1ΔE9. These cells were made either via co-transfection with CSCW-APPSL-GFP and CSCW-PS1ΔE9-mCherry vectors or via single transfection with CSCW-APPSL-PS1ΔE9-mCherry [13, 14]. Depending on the transfection efficiency, repeated transfections are possible.

To enrich the ReN cells with high levels of APPSL-GFP, PS1ΔE9-mCherry, or APPSL-PS1ΔE9-mCherry, we also performed fluorescence-activated cell sorting (FACS) enrichment after the transfection. First, the transfected ReN cells were dislodged as described above. The cell pellets were then resuspended in PBS supplemented with 2% serum replacement solution and 2% B27 and passed through a cell strainer mesh (70 μm Nylon). The concentration was adjusted to 2×10^6 cells per mL and the suspension was enriched via the FACS Aria cell sorter, using the 100 μm nozzle and a sheath pressure of 20 psi (Broad Institute of MIT and Harvard, Cambridge, MA). GFP and/or mCherry channels were used to detect the expression of the transfected genes in the individual cells. Immediately after sorting, the cells were plated on Matrigel-coated dishes.

To culture the FACS-sorted cells, they were first spun down at $2000 \times g$ for 3 min at 4 °C. After the supernatant was removed, the pellet was resuspended in pre-warmed ReN proliferation medium (1 mL per 2×10^6 collected cells). Cells were seeded on Matrigel-coated 24-well plates at an initial density of 2×10^5 cells/cm², after which they were serially expanded into Matrigel-coated 6-well plates, and finally into Matrigel-coated T25 flasks. High seeding density after sorting is pivotal to promote cell proliferation. Additionally, multiple frozen cell stocks can be made at this stage.

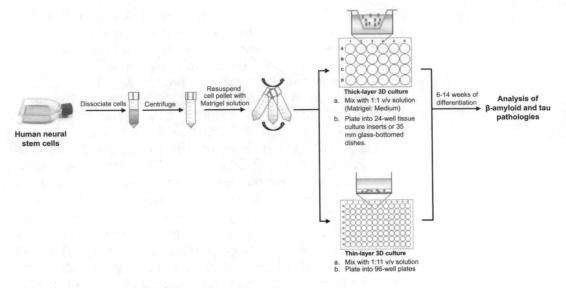

Fig. 1 A brief summary of the ReN cell 3D-culture procedure

Cell passaging is performed as described above. However, it is important to keep track of passage numbers since we observed that several high-expressing cell lines rapidly lost APPSL and PS1ΔE9 expression after >4 passages.

3.2 3D Matrigel Cultures (Thin- and Thick-Layer) and Differentiation

To begin 3D cell culture plating, the ReN cells were first washed and detached according to the procedure described above (Fig. 1). After 3–5 min incubation, the cells were resuspended in 3 mL of ReN differentiation medium. After pipetting the suspension up and down inside the flask at least three times to ensure cell singularization, the suspension was transferred into a 15 mL tube. After centrifuging for 2 min at $2000 \times g$, the supernatant was removed, and the cell pellet resuspended in 2 mL cold ReN differentiation medium and vortexed for 10 s (Fig. 1). For cell counting, a small aliquot of suspended cells was diluted 1:10 (10 μL suspension: 90 μL differentiation medium), and the cell number was measured using a cell counter slide. The optimum concentration is about 2×10^7 cells/mL before adding Matrigel at a 1:1 ratio. We have found that the final concentration of 1×10^7 cells/mL, after adding the Matrigel, is suitable for long-term 3D-differentiation cultures and for detecting robust Aβ and p-tau accumulation by biochemical analysis. The thin-layer (~100–300 μm) and thick-layer (~4 mm) cultures are set up by mixing different concentrations of Matrigel with cells. The thin-layer cultures were employed for immunocytochemistry, and the thick-layers for molecular, biochemical, and ELISA analyses (Figs. 1, 2, and 3).

To generate thick-layer 3D cultures, cold Matrigel was added to the cell suspension on ice [1:1 (vol/vol) dilution] using prechilled   pipette   tips,   then   the   mixture   was   vortexed   for   30   s.   300   μL of the Matrigel–cell suspension was dispensed into each tissue culture insert in 24-well plates using prechilled pipettes (Fig. 1). Plates were then incubated at 37 °C in CO_2 incubator overnight. After 24 h, pre-warmed differentiation medium was added to the plates (1 mL: 500 μL inside of insert and 500 μL outside of insert) and placed in the incubator. Half the volume of medium was changed every 3–4 days from both the inside and outside of the insert. A more frequent media change may be necessary (every 2 days) in the beginning stages, as indicated by the medium acidity. The cells were differentiated for 4–17 weeks, depending on the experimental requirements. Given that thick-layer cultures are not easy to analyze by microscopy, a routine Lactate dehydrogenase (LDH) release assay can be performed in order to monitor the progress, as we have previously described [14].

To set up thin-layer 3D cultures, the 1:1 Matrigel–cell mixture was further diluted. For plating into 96-well plates, we added five additional volumes of cold ReN differentiation medium (1:11 dilution final, Fig. 1). This same dilution ratio was used for plating thin-layer 3D cultures in 8-well/16-well chambered cover glass slides or MatTek glass-bottomed dishes. A small volume of 1:1 Matrigel–cell mixture was directly loaded onto these glass-bottomed dishes, yielding a quasi thin-layer culture used for confocal immunofluorescence analysis. Using prechilled pipettes, 100 μL of the Matrigel–cell suspension mixture was plated per well of the 96-well plates. If a thicker 3D culture is desired, 2 drops per well (160 μL if using a multichannel pipette) can be added. A volume of 200 μL is recommended for 8-well chambered cover glass slides and 300 μL for glass-bottomed dishes. The plates were incubated at 37 °C in a CO_2 incubator overnight. The next day, 2 drops of pre-warmed ReN differentiation medium were added to each well of the 96-well plates (160 μL using a multichannel pipette), 200 μL to the 8-well chambered cover glass slides and 300 μL to the glass-bottomed dishes. Media replacement follows the same guidelines as for the thick-layer cultures described above. Cell differentiation in these thin-layer 3D cultures was closely monitored by optical and fluorescence microscopy. The cells were differentiated for 4–12 weeks, depending on the experiment (Figs. 2 and 3). If needed, the cultures can be fixed with 4% paraformaldehyde (PFA) at 2–4 weeks for testing of neural marker expressions by immunofluorescence staining or immunohistochemistry. To monitor proper 3D differentiation in both the thin and thick cultures, expression of adult neural markers, including MAP2, NeuN, NR2B, vGluT, Synapsin 1, and 4R tau, can be analyzed using RT-PCR, qRT-PCR, and immunofluorescence staining, as described previously (Figs. 2 and 3) [13, 14].

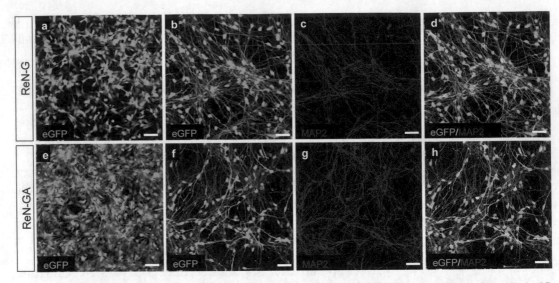

Fig. 2 Immunofluorescence analysis of 2-week differentiated control (ReN-G) and fAD (ReN-GA) cells in 3D cultures. Representative confocal images of undifferentiated (**a** and **e**) or 2-week differentiated (**b–d** and **f–h**) ReN-G (**a–d**) and ReN-GA (**e–h**) cells, labeled with MAP2 (a neuronal marker, *red*). eGFP, *green*. Overlay images are in (**d**) and (**h**). Scale bars, 50 μm

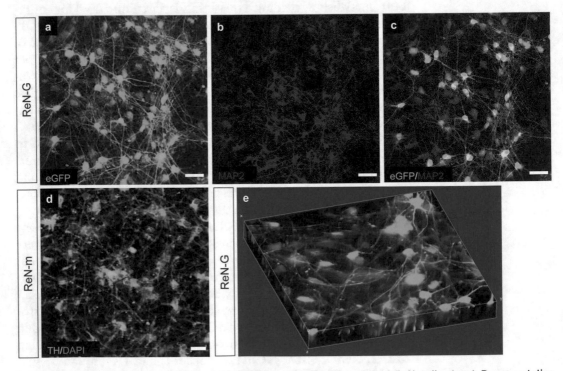

Fig. 3 Immunofluorescence analysis of 12- or 17-week 3D-differentiated ReN cells. (**a–c**) Representative images of 12-week differentiated ReN-G cells labeled with MAP2 (a neuronal marker, *red*). Scale bars, 50 μm. (**d**) Immunofluorescence of 12-week differentiated ReN-m cells stained with TH (tyrosine hydroxylase, *green*). DAPI, *blue*. Scale bars, 20 μm. (**e**) 3D reconstituted image of 17-week differentiated ReN-G cells

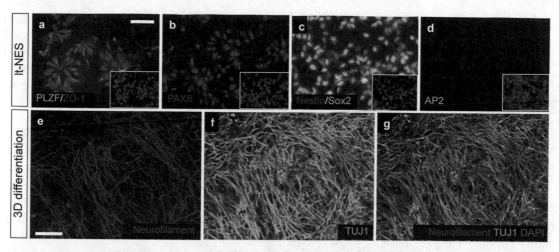

Fig. 4 3D-Neuronal differentiation of lt-NES cells. (**a–d**) lt-NES cells express typical neural precursor markers PLZF (**a**), ZO-1 (**a**), PAX6 (**b**), Nestin (**c**), and Sox2 (**c**). The cells exhibit typical rosette morphology, while no contaminating neural crest cells appear in the AP-2 staining (**d**). Scale bars, 25 μm. (**e–g**) After 2 weeks differentiation in thin-layer 3D-culture, neuronal markers are strongly expressed (neurofilament, (**e**); TUJ1, (**f**); overlay, (**g**)). DAPI, blue. Images of 3D-cultures from ApoTome microscopy. Scale bar, 100 μm

3.3 3D Differentiation of iPSC-Derived lt-NES Cells

The widespread availability of iPSC technology [25] has enabled many new and innovative models for human disorders by providing access to nearly unlimited numbers of human tissue-specific cells. To generate lt-NES cells, iPSCs were first aggregated into free-floating spheres. To that end, colonies were incubated with EDTA for approximately 5 min for easier detachment. After the EDTA was removed, the colonies were washed off with E8 medium containing 10 μM ROCK inhibitor. To facilitate sphere formation, colonies were transferred to uncoated petri dishes. Once spherical aggregates were established, the iPSCs were induced towards a neural fate and processed as described by Koch and colleagues [26]. To validate cellular identity, expression of the neural precursor markers Nestin, Sox2, Pax6, and ZO-1 were confirmed by immunocytochemistry. Typical rosette morphology should be easily visualized by co-staining of Sox2 and ZO-1, with focused ZO-1 immunoreactivity exclusively in the rosette center (Fig. 4). lt-NES cells were dislodged via incubation with 0.05% trypsin in PBS at 37 °C for 5–10 min. The reaction was stopped by the addition of trypsin inhibitor (Life Technologies, Waltham, MA), and the cell suspension was centrifuged, resuspended, and plated on fresh PO/Ln-coated dishes. High culture density is pivotal during the first five to ten passages, therefore splitting ratios should not exceed 1:2.

3D cultures (both thin- and thick-layer) were established in analogy to the ReN cell setup. Lt-NES cells in the exponential growth phase were dislodged with trypsin, pelleted and resuspended in ice-cold lt-NES medium. The cell concentration was adjusted to 1×10^7 cells/mL for thick-layer cultures or 2×10^6 cells/mL for thin-layer cultures. Thick-layer cultures were prepared by mixing 200 μL of cell suspension with 200 μL of Matrigel and plated into a 96-well format transwell insert. For thin-layer cultures, 90 μL of cell suspension were mixed with 10 μL Matrigel and deposited into one well of a clear-bottom 96-well plate. To solidify the Matrigel, freshly prepared cultures were placed into an incubator at 37 °C for 30–60 min. Then, It-NES medium was added to completely fill the wells for thin-layer cultures or the recipient 24-well for thick-layer cultures. After 48 h, the medium was replaced with neural differentiation medium. Equilibrating the medium to 37 °C, 5% CO_2 prior to application to the cells reduced pH spikes. In our experience, medium turnover might appear very high during the first few weeks as indicated by acidification, thus necessitating daily replacement.

After 14 days, thin-layer cultures present various neuronal markers, such as βIII-tubulin (TUJ1) and neurofilament, in immunohistochemical analysis at very high culture densities. These culture conditions lay the foundation for long-term differentiation of iPSC-derived neurons in complex, 3D network structures (Fig. 4).

3.4 Analysis of Aβ and p-tau Pathologies Using Thin-Layer Cultures

To prepare for immunofluorescence and immunohistochemical staining, the thin-layer 3D cultures were fixed with 4% PFA. First, the wells were washed once with D-PBS and then treated with 100 μL of 4% PFA solution per well, allowing for overnight incubation at room temperature. The PFA solution was removed the next day, followed by three rinses with D-PBS. These fixed 3D cultures were then permeabilized by incubating with TBST containing 0.5% Triton X-100 and 4% goat IgG (200 μL per well) at room temperature for 1 h, and blocked by incubating with blocking/dilution solution supplemented with 4% goat IgG (200 μL per well) at 4 °C overnight with gentle rocking.

The following day, the cultures were washed with TBST once for 10 min and incubated with primary antibodies in blocking/dilution solution with 4% goat IgG at 4 °C overnight with gentle rocking. After washing with TBST five times for 10 min each, the cultures were incubated with Alexa Fluor secondary antibodies for 5 h at room temperature with gentle rocking, followed by another five rounds of washing with TBST for 5 min each. Finally, a drop (~20 μL) of anti-fade gold (Life Technologies, Waltham, MA) was added on top of the fixed/stained thin-layer 3D culture sections to avoid fluorescence quenching, and the fluorescence images were captured using an Olympus DSU confocal microscope (Olympus USA, Center Valley, PA) or Nikon A1 confocal laser microscope system (Tokyo, Japan) (Fig. 5). Captured images were analyzed by ImageJ (a public domain image analysis software, NIH), IPlabs (IP Labs, BioVision Technologies, Exton, PA or MetaMorph (Olympus, Molecular Devices, Sunnyvale, CA). Since most of our

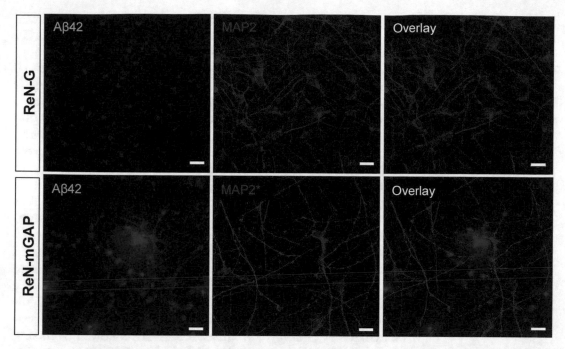

Fig. 5 Immunofluorescence analysis of β-amyloid deposition in control (ReN-G) and FAD (ReN-mGAP) cells after 7-week 3D differentiation. Representative images of 7-week differentiated ReN-G (*upper panels*) and ReN-mGAP (*lower panels*) labeled with anti-Aβ42 antibody, neuronal marker MAP2, and overlay images. *MAP2 staining overlaps with mCherry signal expressed in mGAP cells. Scale bars, 20 μm

FAD ReN cells express high levels of GFP and mCherry, the UV-compatible amyloid dye Amylo-Glo can be used for detecting β-amyloid aggregates in the thin layer 3D culture, as described previously [13, 14].

3.5 Analysis of Aβ Pathology Using Thick-Layer Cultures

We used TBS/2%SDS/Formic acid serial extractions for detecting Aβ aggregates. The accumulation of SDS-resistant Aβ oligomers, including dimers, trimers, and tetramers, can also be used as an indirect marker for Aβ aggregation. To prepare the Matrigel–cell pellet for biochemical analyses from thick-layer 3D cultures, the plate was placed on ice, and the inserts were carefully taken out using forceps. The 3D Matrigel culture was released from the insert by cutting three-fourths along the outer edge of the bottom membrane of the insert and placing the insert halfway into a microcentrifuge tube. They were then centrifuged at $3200 \times g$ (Beckman Coulter, Danvers, MA) for 30 s at room temperature (~24 °C). The insert plastic component was discarded and the culture collected inside the tube was centrifuged again at $8000 \times g$ for 5 min at 4 °C. After careful removal of the medium supernatant, the Matrigel–cell pellet suspension was stored at −80 °C until use (generally, it can be stored at −80 °C for several months).

On the day of experiment, the Matrigel–cell pellet was thawed on ice for 10 min. The same volume of 2× TBS extraction buffer (~100–200 μL in general) was added and the mixture was homogenized on ice using a disposable-tip rotor-driven homogenizer 10 times up and down, at which point the gels became clear/mildly cloudy. They were further sonicated for 10 min at 4 °C using a sonic cleaning water bath, and subsequently spun down at 100,000 × g using an Optima TL Ultracentrifuge (Beckman Coulter, Danvers, MA) for 1 h at 4 °C. The supernatant was collected as the TBS-soluble fraction and, after measuring the protein levels, was frozen at −20 °C. The remaining pellet was homogenized with 50 μL of 1× SDS extraction buffer, following the same protocol as described for the TBS-soluble fraction. The supernatant was saved as the TBS-insoluble/2% SDS-soluble fraction and frozen at −20 °C after measuring protein levels. Lastly, the pellet was washed with 100 μL 1× SDS extraction buffer briefly, and homogenized with 10 μL of 70% formic acid on ice following the same procedure as the preceding extractions. The collected supernatant (TBS-insoluble/2% SDS-insoluble/formic-acid-soluble fractions) was neutralized by adding 10 volumes of 2 M Tris followed by mixing with a quarter volume of 4× LDS buffer (Life Technologies, Waltham, MA) containing 8% β-mercaptoethanol. Protein levels of SDS-soluble fractions were used to normalize the total protein levels in TBS and formic acid fractions.

For Western blot analysis of Aβ, the extracted samples were run on a 12% NuPAGE gel. Proteins were transferred from the SDS-PAGE gels to PVDF membranes. The membrane was incubated with 0.5% glutaraldehyde solution for 10 min, and blocked with SuperBlock blocking solution for 1 h at room temperature. It was then incubated in primary antibody solution (see Sect. 2.3) in 4% BSA at 4 °C overnight with rocking. On the next day, the membrane was washed with TBST three times, incubated in secondary antibody solution for 1 h at room temperature with gentle rocking, and washed again with TBST three times. The blots were developed with SuperSignal Dura or Femto ECL solutions.

3.6 Analysis of p-tau Pathology Using Thick-Layer Cultures

For biochemical analyses of p-tau, we used a 1% Sarkosyl extraction. Utilizing the steps above, after sonicating, the TBS-extracted samples were mixed with an equal volume of 2× RIPA buffer, and homogenized again using a rotor-driven homogenizer on ice. Following 15 min incubation on ice, the samples were sonicated for 10 min, and centrifuged using a tabletop centrifuge at 10,000 × g for 5 min at 4 °C. The supernatant was transferred into a new tube (supernatant A). The remaining pellet was resuspended in 80 μL of 1× RIPA and, after sonication for 10 min, was centrifuged at 10,000 × g with a tabletop centrifuge for 5 min at 4 °C. This supernatant was collected as supernatant B. Supernatants A and B were combined and 1/20th volume of 20% Sarkosyl solution was added, followed by incubation at room temperature for 60 min in a rotary mixer. After centrifugation at $150,000 \times g$ for 1 h at 4 °C, the new supernatant was transferred into fresh tubes (Sarkosyl-soluble fraction) and protein levels were measured. The pellet was washed briefly with 100 μL 2× RIPA buffer, and then three times with 0.5 mL PBS. Next, 1× LDS sample buffer was added with 10 M Urea and 2% β-mercaptoethanol and the solution was heated to 95 °C for 5 min. The supernatant was collected as Sarkosyl-insoluble fraction.

For Western blot, 15–75 mg of proteins were resolved on 12% Bis-Tris or 4–12% gradient Bis/Tris gels (Life Technologies, Waltham, MA). Proteins were transferred from the SDS-PAGE gels to Immun-Blot polyvinylidene fluoride (PVDF) membranes (Bio-Rad, Hercules, CA), and the membrane was blocked with either 4% BSA or SuperBlock blocking solution for 1 h at room temperature. For p-tau blots, skim milk should not be used for blocking since it contains active phosphatase. The membrane was incubated in primary antibody solution (see Sect. 2.3) at 4 °C overnight with rocking. On the next day, the membrane was washed with TBST for 10 min three times, and incubated in secondary antibody solution for 1 h at room temperature with gentle rocking. After washing in TBST for 10 min three times, the blot was developed with SuperSignal Dura or Femto ECL solutions.

For dot blot analysis, the nitrocellulose membrane was wet in 1× TBS buffer for 5 min and located on top of a wet Bio-Rad blotting paper. 0.5–1 μL of protein samples were loaded and the membrane was incubated for 20 min until the protein samples were completely absorbed into the membrane. The membrane was blocked with either 4% BSA or SuperBlock blocking solution for 1 h at room temperature, and incubated in the primary antibody solution (see Sect. 2.3) at 4 °C overnight with rocking. On the next day, the membrane was washed with TBST for 10 min three times and incubated in the secondary antibody solution for 1 h at room temperature with gentle rocking. After washing with TBST for 10 min three times, the blot was developed with SuperSignal Dura or Femto ECL solutions.

4   Conclusion

In this chapter, we describe protocols for establishing and analyzing 3D human neural cell culture systems recapitulating AD pathogenesis including Aβ plaques and tau tangles. The levels and distribution of Aβ and p-tau aggregates depend on the FAD stem cell line passage number, the type of FAD mutations, the number of cells seeded, and the enrichment of high expressing cells. In particular, we observed that the levels of pathogenic Aβ42 are very important to achieve high levels of Aβ aggregates and p-tau pathology. Generally, we observed extracellular Aβ aggregates starting at 6 weeks of differentiation (Fig. 5) and robust p-tau accumulation after 10 weeks [13, 14]. Accumulations of p-tau in cell bodies and neurites were detected by immunostaining with p-tau antibodies after 8 weeks, whereas robust accumulation of p-tau aggregates was detected after 10–14 weeks. We also observed that both Aβ aggregates and p-tau pathology were further elevated through 17 weeks, depending on the cell lines.

Although our Matrigel-based 3D culture system is optimized for ReN cells, we have shown here that human iPSC-derived neural stem cells (It-NES) were successfully cultured and differentiated into mature neurons in our 3D culture system (Fig. 4). This suggests that our 3D culture protocol can be easily adapted to generate models of various neurological diseases, including AD, by differentiating human patient-specific iPSC-derived neuronal cells. In addition to Aβ aggregation, we have shown that our 3D culture conditions not only enhance Aβ aggregation, but also accelerate neuronal maturation at least for the ReN cells (Fig. 4) [13, 14].

Given the advantages of this 3D cell culture system, various applications of 3D cell models have been under investigation. The most valuable application of 3D neural cell culture models is a large-scale high-throughput screening for novel therapeutic targets, and validation of targets during the initial stages of drug discovery, which has not been feasible in the current mouse models [8, 13, 14, 27]. Human stem cell-derived models provide a valid system to test the efficacy and potential toxicity of candidate AD drugs. Cross-checking a candidate drug target in both human- and mouse-based models would minimize the chance of failure in the final stages of clinical trials [28]. Another key application of 3D neural cell culture models of AD is the investigation into the molecular mechanisms underlying the AD pathology. This will provide novel drug targets and may lead to the discovery of new diagnostic and prognostic biomarkers of AD [29]. The 3D neural cell culture platform of AD might also be an ideal model to test the impact of AD-associated genetic variants and environmental factors on AD pathogenesis.

Acknowledgments

This work is supported by grants from the Cure Alzheimer's fund to D. Y. K., S. H. C., and R. E. T., the Bio & Medical Technology Development Program of the National Research Foundation (funded by the Korean government, MSIP (2015M3A9C7030151, Y.H.K)), and National Institutes of Health (1RF1AG048080-01, D.Y.K. and R.E.T.; 5P01AG15379, D.Y.K. and R.E.T.; 2R01AG014713, D.Y.K.; 5R37MH060009, R.E.T.). We would also like to thank Mrs. Jenna Aronson (MGH) and Mr. Kevin Washicosky (MGH) for technical help in producing figures.

References

1. Tanzi RE, Bertram L (2005) Twenty years of the Alzheimer's disease amyloid hypothesis: a genetic perspective. Cell 120(4):545–555. doi:10.1016/j.cell.2005.02.008
2. D'Avanzo C, Aronson J, Kim YH, Choi SH, Tanzi RE, Kim DY (2015) Alzheimer's in 3D culture: Challenges and perspectives. Bioessays. doi:10.1002/bies.201500063
3. Lewis J, Dickson DW, Lin WL, Chisholm L, Corral A, Jones G, Yen SH, Sahara N, Skipper L, Yager D, Eckman C, Hardy J, Hutton M, McGowan E (2001) Enhanced neurofibrillary degeneration in transgenic mice expressing mutant tau and APP. Science 293(5534):1487–1491. doi:10.1126/science.1058189
4. Oddo S, Caccamo A, Shepherd JD, Murphy MP, Golde TE, Kayed R, Metherate R, Mattson MP, Akbari Y, LaFerla FM (2003) Triple-transgenic model of Alzheimer's disease with plaques and tangles: intracellular Abeta and synaptic dysfunction. Neuron 39(3):409–421
5. Duan L, Bhattacharyya BJ, Belmadani A, Pan L, Miller RJ, Kessler JA (2014) Stem cell derived basal forebrain cholinergic neurons from Alzheimer's disease patients are more susceptible to cell death. Mol Neurodegener 9(1):3. doi:10.1186/1750-1326-9-3
6. Israel MA, Yuan SH, Bardy C, Reyna SM, Mu Y, Herrera C, Hefferan MP, Van Gorp S, Nazor KL, Boscolo FS, Carson CT, Laurent LC, Marsala M, Gage FH, Remes AM, Koo EH, Goldstein LSB (2012) Probing sporadic and familial Alzheimer's disease using induced pluripotent stem cells. Nature. doi:10.1038/nature10821
7. Kondo T, Asai M, Tsukita K, Kutoku Y, Ohsawa Y, Sunada Y, Imamura K, Egawa N, Yahata N, Okita K, Takahashi K, Asaka I, Aoi T, Watanabe A, Watanabe K, Kadoya C, Nakano R, Watanabe D, Maruyama K, Hori O, Hibino S, Choshi T, Nakahata T, Hioki H, Kaneko T, Naitoh M, Yoshikawa K, Yamawaki S, Suzuki S, Hata R, S-i U, Seki T, Kobayashi K, Toda T, Murakami K, Irie K, Klein WL, Mori H, Asada T, Takahashi R, Iwata N, Yamanaka S, Inoue H (2013) Modeling Alzheimer's disease with iPSCs reveals stress phenotypes associated with intracellular Aβ and differential drug responsiveness. Cell Stem Cell 12(4):487–496. doi:10.1016/j.stem.2013.01.009
8. Mohamet L (2014) Familial Alzheimer's disease modelling using induced pluripotent stem cell technology. World J Stem Cells 6(2):239. doi:10.4252/wjsc.v6.i2.239
9. Moore S, Evans LDB, Andersson T, Portelius E, Smith J, Dias TB, Saurat N, McGlade A, Kirwan P, Blennow K, Hardy J, Zetterberg H, Livesey FJ (2015) APP metabolism regulates tau proteostasis in human cerebral cortex neurons. Cell Rep 11(5):689–696. doi:10.1016/j.celrep.2015.03.068
10. Muratore CR, Rice HC, Srikanth P, Callahan DG, Shin T, Benjamin LNP, Walsh DM, Selkoe DJ, Young-Pearse TL (2014) The familial Alzheimer's disease APPV717I mutation alters APP processing and Tau expression in iPSC-derived neurons. Hum Mol Genet 23(13):3523–3536. doi:10.1093/hmg/ddu064
11. Sproul AA, Jacob S, Pre D, Kim SH, Nestor MW, Navarro-Sobrino M, Santa-Maria I, Zimmer M, Aubry S, Steele JW, Kahler DJ, Dranovsky A, Arancio O, Crary JF, Gandy S, Noggle SA (2014) Characterization and molecular profiling of PSEN1 familial Alzheimer's disease iPSC-derived neural progenitors. PLoS One 9(1):e84547. doi:10.1371/journal.pone.0084547
12. Yagi T, Ito D, Okada Y, Akamatsu W, Nihei Y, Yoshizaki T, Yamanaka S, Okano H, Suzuki N (2011) Modeling familial Alzheimer's disease with induced pluripotent stem cells. Hum Mol Genet 20(23):4530–4539. doi:10.1093/hmg/ddr394
13. Choi SH, Kim YH, Hebisch M, Sliwinski C, Lee S, D'Avanzo C, Chen H, Hooli B, Asselin C, Muffat J, Klee JB, Zhang C, Wainger BJ, Peitz M, Kovacs DM, Woolf CJ, Wagner SL, Tanzi RE, Kim DY (2014) A three-dimensional human neural cell culture model of Alzheimer's disease. Nature 515(7526):274–278. doi:10.1038/nature13800
14. Kim YH, Choi SH, D'Avanzo C, Hebisch M, Sliwinski C, Bylykbashi E, Washicosky KJ, Klee JB, Brüstle O, Tanzi RE, Kim DY (2015) A 3D human neural cell culture system for modeling Alzheimer's disease. Nat Protoc 10(7):985–1006. doi:10.1038/nprot.2015.065
15. Haycock JW (2010) 3D cell culture: a review of current approaches and techniques. In: Amyloid proteins, vol 695. Humana Press, Totowa, NJ, pp 1–15. doi:10.1007/978-1-60761-984-0_1
16. LaPlaca MC, Vernekar VN, Shoemaker JT, Cullen DK (2010) Three-dimensional neuronal cultures. In: Morgan JR, Berthiaume F (eds) Methods in bioengineering: 3D tissue engineering, vol 11, pp 187–204
17. Li H, Wijekoon A, Leipzig ND (2012) 3D differentiation of neural stem cells in macroporous photopolymerizable hydrogel scaffolds. PLoS One 7(11):e48824. doi:10.1371/journal.pone.0048824

18. Liedmann A, Frech S, Morgan PJ, Rolfs A, Frech MJ (2012) Differentiation of human neural progenitor cells in functionalized hydrogel matrices. Biores Open Access 1(1):16–24. doi:10.1089/biores.2012.0209

19. Hughes CS, Postovit LM, Lajoie GA (2010) Matrigel: a complex protein mixture required for optimal growth of cell culture. Proteomics 10(9):1886–1890. doi:10.1002/pmic.200900758

20. Liedmann A, Rolfs A, Frech MJ (2012) Cultivation of human neural progenitor cells in a 3-dimensional self-assembling peptide hydrogel. J Vis Exp 59. doi:10.3791/3830

21. Ortinau S, Schmich J, Block S, Liedmann A, Jonas L, Weiss DG, Helm CA, Rolfs A, Frech MJ (2010) Effect of 3D-scaffold formation on differentiation and survival in human neural progenitor cells. Biomed Eng Online 9(1):70. doi:10.1186/1475-925X-9-70

22. Baker BM, Chen CS (2012) Deconstructing the third dimension—how 3D culture microenvironments alter cellular cues. J Cell Sci 125(13):3015–3024. doi:10.1242/jcs.079509

23. Chen G, Gulbranson DR, Hou Z, Bolin JM, Ruotti V, Probasco MD, Smuga-Otto K, Howden SE, Diol NR, Propson NE, Wagner R, Lee GO, Antosiewicz-Bourget J, Teng JM, Thomson JA (2011) Chemically defined conditions for human iPSC derivation and culture. Nat Methods 8(5):424–429. doi:10.1038/nmeth.1593

24. Beers J, Gulbranson DR, George N, Siniscalchi LI, Jones J, Thomson JA, Chen G (2012) Passaging and colony expansion of human pluripotent stem cells by enzyme-free dissociation in chemically defined culture conditions. Nat Protoc 7(11):2029–2040. doi:10.1038/nprot.2012.130

25. Takahashi K, Tanabe K, Ohnuki M, Narita M, Ichisaka T, Tomoda K, Yamanaka S (2007) Induction of pluripotent stem cells from adult human fibroblasts by defined factors. Cell 131(5):861–872. doi:10.1016/j.cell.2007.11.019

26. Koch P, Opitz T, Steinbeck JA, Ladewig J, Brustle O (2009) A rosette-type, self-renewing human ES cell-derived neural stem cell with potential for in vitro instruction and synaptic integration. Proc Natl Acad Sci U S A 106(9):3225–3230. doi:10.1073/pnas.0808387106

27. Young W (2012) Patient-specific induced pluripotent stem cells as a platform for disease modeling, drug discovery and precision personalized medicine. J Stem Cell Res Ther S10:010. doi:10.4172/2157-7633.S10-010

28. De Strooper B (2014) Lessons from a failed γ-Secretase Alzheimer trial. Cell 159(4):721–726. doi:10.1016/j.cell.2014.10.016

29. Livesey FJ (2014) Human stem cell models of dementia. Hum Mol Genet 23(R1):R35–R39. doi:10.1093/hmg/ddu302

Chapter 2

Neural Stem Cell Fate Control on Micropatterned Substrates

Leonora Buzanska, Marzena Zychowicz, Ana Ruiz, and François Rossi

Abstract

Neural stem cell fate decisions are dependent upon signals coming from microenvironment composed of extracellular matrix proteins, soluble factors and specificity of cell–cell contacts. We have developed in vitro systems to trace developmental processes of neural stem cells and to control their fate commitment. Applied technologies include nano/micro-fabrication techniques, like microcontact printing and piezoelectric microspotting of biomolecules on plasma deposited cell repellent surface. Designed bioactive domains with controlled content and geometry served as a template to immobilize neural stem cells to the surface and direct their differentiation. Migration and axon-like outgrowth have been successfully guided by means of interconnected squares configuration. Receptor mediated versus electrostatic interactions on the cell membrane–surface interface were crucial to keep the cells either in neurally committed or in non-differentiated stages by fibronectin or poly-L-lysine pattern, respectively. Single cell versus multicellular positioning further promoted stem cells non-differentiated stage. Activation of intracellular pathways by signaling molecules (Wnt, CNTF, Jagged, Notch and DKK-1) microspotted with fibronectin directed differentiation into astrocytic and neuronal lineages, as revealed by immunocytochemical and molecular analysis. Our results proved that neural stem cell fate decisions can be influenced by manipulating the composition and architecture of the 2D bioactive domains reflecting their natural niche microenvironment.

Key words Neural stem cells, Piezoelectric microspotting, Microcontact printing, Bioactive domains

1 Introduction

Human neural stem cells are widely investigated due to their promise held for the therapy of neurological disorders. Their specific properties to incorporate in vivo into the damaged neuronal tissue and differentiate into specific neural cell types, favor the opinion of their possible replacement rather than adjuvant features, as compared to mesenchymal stem cells. Thus, it is important to set up optimized in vitro microenvironments mimicking in vivo conditions and enabling to control human Neural Stem Cells (hNSC) fate commitment and differentiation. Standardized conditions for in vitro culture and fate commitment of therapeutically competent hNSC are, although widely studied, not thor-

Amit K. Srivastava et al. (eds.), *Stem Cell Technologies in Neuroscience*, Neuromethods, vol. 126,
DOI 10.1007/978-1-4939-7024-7_2, © Springer Science+Business Media LLC 2017

oughly characterized. Biotechnological innovations provide tools, such as advanced stem cell culture protocols and generation of new, tunable biomaterials capable of releasing specific signals in a spatially and temporally controlled manner. These innovations together with emerging technologies of fabrication of 2D and 3D cell culture platforms provide opportunities to create in vitro nature-like conditions for stem cell-based systems. Such in vitro systems are referred to, in the literature, as "biomimetic" due to their ability to mimic in vivo native microenvironments [1, 2]. To adequately mimic in vivo neural stem cell niche, in vitro one should learn from the experience and knowledge of in vivo systems. The fate of neural stem cells is controlled by a number of signaling cues present in the stem cell niche. In addition to structural elements like extracellular matrix (ECM) proteins, proteoglycans and other cells, microenvironment of the niche also include soluble factors, such as cytokines, growth factors, cell adhesion molecules, hormones, and neurotransmitters. Complex mutually time- and space-dependent cell–cell and cell–ECM protein/proteoglycan interactions have been shown to play an essential role in cell fate decisions [3, 4]. These interactions may influence the binding of soluble signaling molecules to cellular receptors and may change their expression [5]. On the other hand, the presence of signaling proteins that bind to their specific receptors, direct cell developmental decisions whether to proliferate and maintain their stemness, or to induce neural differentiation [6]. In vitro culture platforms including structural and soluble elements typical for NSC microenvironments have been proposed to investigate the influence of physical cues of NSC fate decisions, but also to define molecular mechanisms underlying these decisions. Variety of tested physical cues comprised geometry of 2D micropatterns [7–11], substrate stiffness [12, 13], topography of the surface [4, 10] and gradients of substrates in microfluidic systems [14, 15]. Directed differentiation of stem cells by changing nanoscale topography of the adhesive substrates has been linked to the modulated genomic expression and the presence of so called "contact guidance spots" in the cell membrane [16]. Further, the surface-printed microdot array in nanoscale was proposed for controlling axon branch formation [17]. Changing geometry of the pattern of microprinted fibronectin and poly-L-lysine biofunctional domains applied in our experiments reinforced directed neuronal differentiation and axonal outgrowth of Human Umbilical Cord Blood derived Neural Stem Cells (HUCB-NSC) [9, 18]. In addition to cell/microenvironment structural interactions, microengineering techniques were used to trace specific, receptor mediated cellular responses to soluble cues. Independent research groups, including our laboratory, have shown that signaling molecules immobilized to the surface together with ECM proteins influence human neural stem cells fate decisions. Soen et al. [19] using ECM proteins combined with different signaling molecules, applied a combinatory approach for human NSC lineage commitment and successfully directed their differentiation. In our studies, defined signaling molecules (CNTF, Notch, Jagged, Wnt3a, Shh, Dkk-1, BMP4), each spotted as a single domain in repetition with the ECM protein (fibronectin), provided specific cues to fate commitment of HUCB-NSC, either stemness maintenance or differentiation into neuronal or astrocytic lineages [8].

In this report, we introduce bioengineering strategies to build up microenvironments at microscale, for the effective fate control of neural stem cells derived from human cord blood. The single or multi cell microarrays are good alternative for the standard culture systems where cell response to a different adhesive materials and/or bioactive compound is investigated. Moreover, micropatterning provides the conditions where exactly the same culture settings (cell density, medium composition, temperature, humidity, and oxygen concentration) can be obtained within the same culture dish. The bioactive microdomains on cell culture platforms are fabricated using either soft-lithographic technique—microcontact printing or piezoelectric spotting, the latter technique enabling deposition of different microdroplets with defined substrate content.

2   Materials

2.1 Neural Stem Cell Culture

Since the availability of human neural stem cells is very limited, we have established an untransformed cell line that is derived from human umbilical cord blood non-hematopoietic mononuclear cells, which were maintained and selected in the presence of 10% FBS and EGF. These cells, named Human Umbilical Cord Blood Neural Stem Cells (HUCB-NSCs) have molecular and functional properties of neural stem cells: they are clonogenic, and after differentiation give rise to three types of cells: neurons, astrocytes and oligodendroglial cells. This cell line has two population of growth: when cultured in serum free conditions cells are free floating, exhibit undifferentiated state of NSC while when they are grown in the presence of 2% serum and ITS supplement (insulin, transferrin, selenium) cells start to adhere and spontaneously differentiate, exhibiting the neuronal (β-tub III, Map-2, NF200) or glial (GFAP, S100β, NG2) markers [20].

For the purpose of the described experiments cells were maintained in serum free medium (DMEM/F12, B27 1:50, EGF 20 ng/mL, AAS) or low serum medium (DMEM/F12, 2% FBS, ITS, AAS). Cells were passaged once a week at 1:4 ratio and maintained at 37 °C, 5% CO_2, and 95% humidity.

2.2 Cell Culture Platforms with Functional Domains Biopatterned by Microcontact Printing and Microspotting

2.2.1 Substrate Preparation by Plasma Polymerization

Plasma polymerization allows depositing conformal thin polymeric films that can either promote or prevent cell adhesion, depending on the polymer chemical function. Plasma polymers are deposited using a glow discharge created from monomer vapor by using capacitive coupled plasma sources. The glow discharge is generated by a high frequency electric field applied between two parallel plate electrodes and contains ions, free electrons and monomer fragments. The properties of the films can be controlled by tuning the discharge parameters, particularly pressure, gas residence time and power.

Cell repellent polyethylene oxide (PEO) has been used as substrate for protein and cell patterning. The PEO films have been deposited by plasma-enhanced chemical vapor deposition in a capacitive coupled reactor, using a glow discharge in diethylene glycol dimethyl ether vapor (Sigma-Aldrich). The plasma reactor had a vessel size of $300 \times 300 \times 150$ mm and two symmetric internal parallel-plate electrodes of 140 mm diameter separated a distance of 50 mm. The plasma was generated by a radio frequency generator operating at 13.56 MHz and connected to the upper electrode, whereas the bottom electrode was grounded and used as sample holder. A 20 mm -thick PEO-like layer was deposited using a pulsed plasma discharge (time on = 10 ms, time off = 100 ms, nominal power = 2 W) of pure diethylene glycol dimethyl ether vapors and a constant working pressure of 20 mTorr. For SPR imaging analysis, the film deposited on the SPR slide was thinner than 20 nm in order to remain in the thickness range usable by the instrument. Ellipsometry measurements of the same PEO-like film deposited on silicon gave a film thickness of 11.5 nm [21].

Previous studies [22] have demonstrated that pulsed plasma polymerization of pure diethylene glycol dimethyl ether monomer allowed the fabrication of stable thin layers with a high retention of polyethylene oxide character and a strong anti-fouling character. Such films could even be microstructured with other polymer functionalities by using photolithographic processes without altering their surface properties [23].

For microspotting, the films were deposited on optical glass slides sonicated successively in trichloroethane, acetone, and ethanol, 5 min each time, and finally rinsed in ultrapure Millipore water and dried under a N_2 flow. In the case of microcontact printing, the PEO-like layer was deposited directly at the bottom of sterile petri dishes rinsed in pure water.

2.2.2 Microcontact Printing (MCP) on PEO-Like Films

Master Fabrication

The first step of the microcontact printing process is the creation of masters having the geometries of interest to reproduce. Photolithography and plasma etching techniques were used to fabricate silicon masters with depth features of 2.5 μm. A positive photoresist (Microposit S1813, Shipley) was spin coated at 900 rpm for 13 s and 2000 rpm for 1 min on silicon and pre-baked (110 °C, 2 min). The resist was then exposed to a laser beam (λ = 405 nm) through ink-printed slides masks. The exposed photoresist was then developed by immersing the slides in Microposit MF-321 developer (Shipley) for 45 s. The samples were washed thoroughly in water and post baked again at 110 °C for 2 min. The patterns created in the photoresist were transferred to the silicon underneath by reactive ion etching. The pattern transfer was done in a inductively coupled plasma equipment introducing SF_6 gas as reagent. The pressure in the reactor chamber was 10 mTorr, the power was set to 400 W and the bias voltage to −60 V. After plasma etching, the photoresist was dissolved by ultrasonicating the slides in acetone for 1 min. In order to avoid adhesion of the stamp to the master and make the peel-off easier during the replica molding step, the silicon masters were coated with a Teflon-like layer, CFx, deposited by plasma polymerization from C_4F_8 gas at 50 mTorr and −40 V of bias. The presence of the low adhesion teflon-like layer was crucial to achieving a good, undamaged replica of the microstructures by facilitating the PDMS peel off.

Replica Molding

The rubber stamps were fabricated by casting the silicon master in polydimethylsiloxane (PDMS). A mixture of polymer precursors was done vigorously stirring 1:10 parts of prepolymer and curing agent (Sylgard 184). The viscous mixture was poured on the masters, left to degas at room temperature for 1 h and cured at 65 °C for 4 h.

Direct Printing of Biomolecules

The microstructured PDMS stamps, fabricated by casting the silicon masters produced by photolithography, have been used for the direct printing of several biomolecules onto plasma polymerized PEO-like films. The PDMS stamps were ultrasonicated in ethanol for 5 min and cleaned in soft O_2 plasma (200 W, 1.2 mTorr) for 30 s in order to increase the hydrophilic character of the PDMS by forming more silica groups on the surface. The hydrophilic PDMS stamps were then inked at room temperature with the biomolecule solution. Proteins, such as BSA, poly-L-lysine, (PLL) and fibronectin (FN), as well as protein antibodies can be successfully printed [24]. Of particular interest for HUCB-NSC fate control were the PLL and FN patterns. For microcontact printing, FITC-labeled PLL was diluted in carbonate buffer (100 mM) at 25 µg/mL and pH 8.4, while FN from human plasma (Sigma, F0895) was used at concentrations of 42 µg/mL and 167 µg/mL diluted in printing buffer (100 mM acetate (Riedel deHaën), pH 5, 5 mM EDTA (Merck), 0.01% Triton X-100 (Fluka), and 0.1% glycerol (Carlo Erba). Figure 1 shows fluorescence microscopy pictures of PLL patterns, evidencing those different geometries, i.e., size, shape, and separation of PLL domains, can be obtained by tailoring the stamp motifs. PLL patterns at highest resolution (one cell, 10 µm) are also shown. Fluorescent imaging of proteins was carried out in a Zeiss inverted microscope Axiovert 200. The inking time was varied according to the biomolecule to adsorb. PLL was incubated

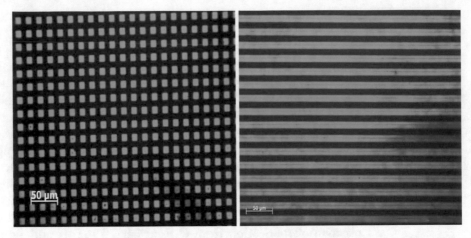

Fig. 1 Fluorescence microscopy images of poly-L-lysine (FITC PLL in *green*) patterns on plasma PEO-like surfaces. The PLL patterns are 10 × 10 μm, *squares* and 10 μm *wide lines*, which commensurate with cell size

for 15 min while fibronectin needed 45 min of incubation to be adsorbed to PDMS. Then, the excess of protein solution was removed; the stamps were dried in a N_2 stream and then put in conformal contact with the PEO-like coated substrates. The PDMS stamp was lifted off vertically and the samples were rinsed with the buffer solution and dried in a nitrogen stream. Before seeding the cells, the patterned substrates were UV sterilized for 15 min. Using this method, the biomolecules attached to the PDMS can transfer and remain adhering to the substrate [7, 21].

2.2.3 Microspotting on PEO-Like Films

In order to obtain the custom dependent array of different molecules on the same platform we have used the microspotting technology. This allows placing the desired biomolecules in designed manner, which means the size of domains, the layout of array, different compositions and concentrations of spotted solution, spatial distribution of domains on the cultured surface, precise volume of spotted solution, making a very reproducible array. This technology, as almost all micropatterning techniques, requires the cell-repellent surface to be deposited on the culture area in order to achieve the fouling/non-fouling contrast. One method to create such non-fouling surface is to deposit a polyethylene oxide-like, PEO-like film, by plasma polymerization [25, 26]. It is a fast method and can be operated on plastic, glass, also within the chambers or standard culture vessels (petri dish) [25]. Moreover, when the PEO-like surface is dry it provides very good immobilization of microspotted proteins.

Spotting of the proteins on the substrates was performed in a S3 sciFLEXARRAYER (Scienion AG, Germany) piezoelectric dispenser. The instrument has a three axis micro-positioning system (accuracy 10 mm) and is equipped with a glass nozzle 80 μm in diameter. A stroboscopic camera allows visual monitoring to adjust piezo voltages and pulse durations for reliable droplet ejection and to avoid satellite drops. Single drops ejected from the nozzle have a volume of 400 pl. Between 1- and 5-drop volumes were dispensed to obtain surface domains of different sizes. The vertical separation between the nozzle and the substrate was typically 0.5 mm. The distance between spots can vary but typically for cell culture experiments a pitch of 600 μm was used. Four-drop volume spots and a pitch of 600 μm have been chosen to prepare microarrays of different biomolecules. They included fibronectin, laminin, collagen I, collagen III, collagen V, bovine serum albumin (BSA), and poly-L-lysine. Finally, fibronectin was chosen as optimal for HUCB-NSC adhesion and arrays of 10×10 spots arranged in two lines of three arrays each were prepared on half slides using FN diluted in printing buffer at concentrations of 21, 42, 84, 167, and 333 μg/mL.

3 Methods

3.1 Functional Analysis of Microprinted and Microspotted Domains

3.1.1 ToF–SIMS Mapping of Biomolecules Patterned on PEO-Like Films by MCP

Time of flight—secondary ion mass spectrometry (ToF–SIMS) was used to characterize the protein or polypeptide patterns. ToF–SIMS was done in a TOF-IV spectrometer (ION TOF GmbH, Germany). Figure 2 depicts ToF–SIMS analyses of fibronectin on PEO-like films. The CH_3O mass in the central image of the figure is ascribed to the polyethylene oxide, whereas C–N mass forming bright squares in the right image belongs to the protein. The protein patterns exhibited a firm stability when put in solution, as confirmed by ToF–SIMS analyses evidencing the presence of well-defined impressions after 24 h in water.

Field of view: 500.0×500.0 μm²

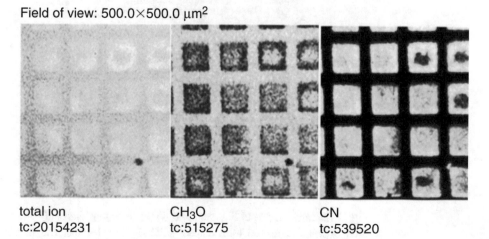

total ion	CH₃O	CN
tc:20154231	tc:515275	tc:539520

Fig. 2 ToF–SIMS mapping of Fibronectin patterned on PEO-like films by MCP

3.1.2 Ellipsometry of Functional Domains

The methodology of ellipsometry analysis has been described in detail in (Ceriotti et al 2007, Ruiz 2013). Ellipsometric data were acquired with a variable angle imaging ellipsometer (model EP3 by Nanofilm Surface Analysis GmbH, Germany) and present as the thickness map of the biomolecule. All imaging measurements were performed in air at room temperature at an angle of incidence of 42°, using a monochromatized high power Xe lamp at a wavelength of $\lambda = 554.3$ nm. A polarizer-compensator sample- analyzer (PCSA) null-ellipsometric procedure was used to obtain maps of the Δ and Ψ angles for the selected area. Thickness maps were calculated from the Δ and Ψ maps by point-by-point modeling with the software EP3View provided with the ellipsometer, using a two-layer model with PEO as first layer and the stamped FN as the second. The thickness (11.5 nm) and the refractive index of plasma deposited PEO ($nPEO = 1.52$) were independently determined by an angle-resolved measurement. A refractive index of 1.46 was used for both FN and PLL [7, 27]. Ellipsometric data were acquired with a variable angle imaging ellipsometer and present as the thickness map of the biomolecule. We have shown equal distribution of protein deposited by MCP (Figs. 3 and 4), and for fibronectin microspotted at different protein concentration, (Fig. 5).

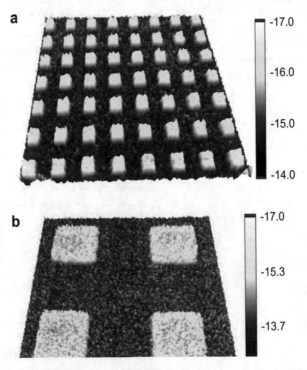

Fig. 3 Thickness maps of PLL stamped on plasma-deposited PEO-like at pH 7, after rinsing, calculated from ellipsometric D- and J-maps. The thickness is given in nm with respect to the Si surface. Size of the maps: (**a**) 1500 mm, 1690 mm and (**b**) 380 mm, 390 mm. Adapted from [7]

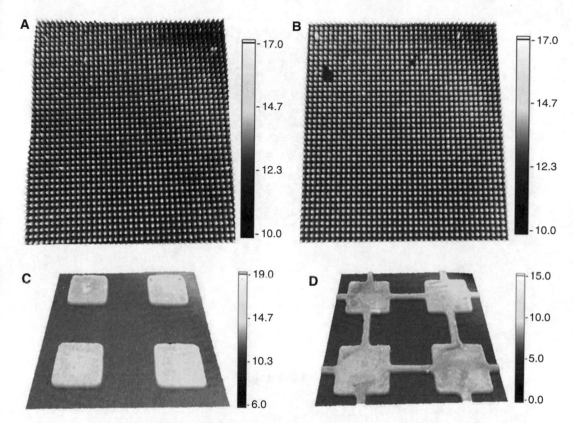

Fig. 4 3D thickness maps of 5 μm *square* MCP patterns of FN 42 μg/mL (**a**) and FN 167 μg/mL (**b**), as well as of 120 μm *square patterns* of FN 42 μg/mL without (**c**) and with (**d**) *interconnecting lines*. The thickness is calculated from ellipsometric Δ- and Ψ-maps and given in nm with respect to the Si surface. Size of the maps: 400 × 400 μm. Adapted from [7]

The amount of FN that remained on the PEO surface after rinsing (Fig. 5b) saturates at $112 \, ng/cm^2$. Using a nominal FN concentration of 84 μg/mL resulted in a retention on PEO close to the saturation value (row 2 and 3 on Fig. 5); thus, the FN concentration of 84 μg/mL was chosen for further spotting experiments.

3.1.3 Surface Plasmon Resonance (SPR) of Biofunctional Domains

Biological functionality of the bioengineered surfaces was characterized by Surface Plasmon Resonance (SPR) to show protein spot reactivity. SPR enabled verification on fibronectin domains of accessibility of the fibronectin receptor binding sites for cell integrin receptors as estimated by binding of anti RGD antibody (Fig. 6).

The reaction of anti-fibronectin monoclonal antibody RGD-specific (Ab-RGD) on spots of six different fibronectin concentrations was tested. We have shown dependence of reactivity of antibody binding through RGD domain upon deposited protein concentration: the reaction was significant starting from the concentration of 42 μg/mL of spotted fibronectin. The control reactions with BSA and anti-ovalbumin were negative (Fig. 6). Thus we confirmed the accessibility of the protein for cell integrin receptor binding.

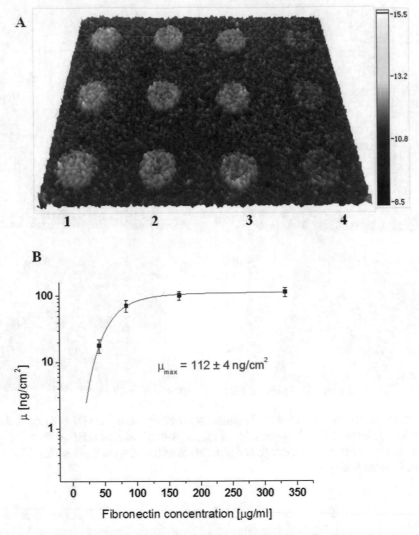

Fig. 5 (**a**) Thickness maps of Fn spots on plasma-deposited PEO-like film after rinsing in water and calculated from ellipsometric D and J maps. The thickness is given in nm with respect to the Si surface and includes the PEO-like layer (11.5 nm). Fibronectin concentration = 167 µg/mL (column 1), 84 µg/mL (columns 2 and 3), and 42 µg/mL (column 4). Size of the maps: 2000–1400 µm. (**b**) Mass density (m) of the fibronectin spots after rinsing as a function of the protein concentration in the spotted solution. It was averaged over six equivalent spots deposited from a solution with equal protein concentration. Adapted from [25]

3.2 Neural Stem Cell Patterning on Microprinted Arrays

Diverse geometry and resolution of the microcontact printed domains can be obtained to accommodate either single cells (pattern size of 10–20 µm) or groups of cells (pattern sizes typically above 100 µm) (Fig. 7). Such arrays were used to study different developmental processes of HUCB-NSC and their responses to the external stimuli. Designed bioactive domains with controlled biomaterial content and geometry served as a template to immobilize neural stem cells to the surface. HUCB-NSC readily adhered to bioactive surface of both: poly-L-lysine and fibronectin domains and presented good confinement to the domains through the 21 days of experiment.

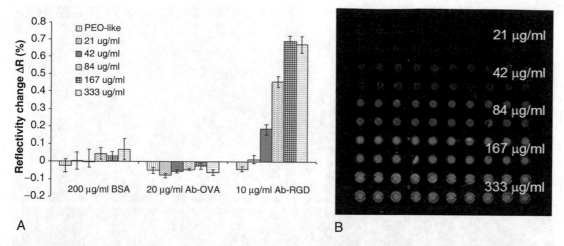

A   B

Fig. 6 (**a**) Reflectivity change due to the injection of 200 mg/mL BSA, 20 µg/mL anti-ovalbumin (Ab-OVA) and 10 µg/mL mouse anti-fibronectin monoclonal antibody RGD-specific (Ab-RGD) on spots of different fibronectin concentrations. (**b**) SPR difference image of the fibronectin array after the interaction with the 10 mg/mL Ab-RGD. Adapted from [25]

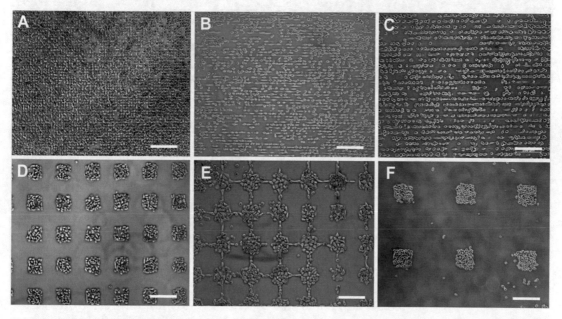

Fig. 7 Patterns of HUCB-NSC grown on PLL: 10 µm posts (**a**), 10 µm-width *lines* (**b**), 20 µm *lines* (**c**), 120 µm *squares* (**d**), 120 µm *squares* and *interconnecting lines* (**e**), and 150 µm *squares* with long distance (**f**), scale bar 200 µm. Adapted from [8]

When using microcontact printing to produce poly-L-lysine and fibronectin patterns of length-scales that commensurate with cell size, single stem cells arrays were achieved and the stem cells were constrained to redistribute their cytoplasm in the permitted areas. Figure 8 shows stem cells patterns on PLL and FN bioactive areas of 10 µm lateral dimensions, either squares or lines, separated by 10 µm cell-repellent gap. While single cell arrays of spherical HUCB-NSC were achieved on the square patterns, the

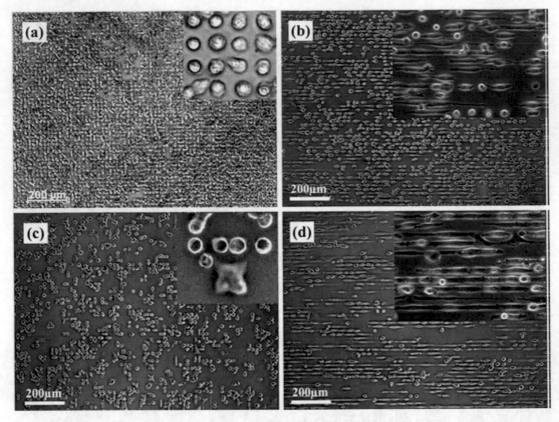

Fig. 8 Phase-contrast images recorded on live cells of HUCB-NSC after 1 day culturing on PLL (**a**, **b**) and fibronectin (**c**, **d**) patterns of 10 × 10 μm *dots* (a, c) and10 μm *wide lines* (**b**, **d**). Adapted from [28]

cells stretched and elongated along the line patterns. The extent of elongation was higher on the FN line pattern (Fig. 8d) than on the PLL one (Fig. 8b), evidencing the dependence of the phenotype on the mode of cell attachment (specific, integrin-mediated versus nonspecific, electrostatic cell binding) [28]. The larger aspect-ratio of the cells grown on fibronectin lines is in agreement with other observations of higher rate of spreading of HUCB-NSC on printed fibronectin as compared to PLL [26]. This is also confirmed by HUCB-NSC scanning electron microscopy images of cells plated either on PLL (Fig. 9a) or on fibronectin (Fig. 9b).

HUCB-NSC attach weakly to PLL through anionic–cationic interactions. The low cell spreading on PLL, together with the background passivation by plasma-PEO, enable obtaining high-density arrays of individual cells, as observed in Fig. 8a. DAPI staining revealed that the nuclei of the cells attached to the 10 μm wide PLL squares occupy all the adhesive PLL area, thus the cells are constrained to maintain a spherical shape. As previously mentioned, cell adhesion to fibronectin involves integrin and focal

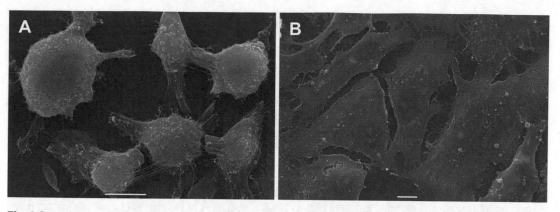

Fig. 9 Scanning electron microscopy images of HUCB-NSC cells plated on poly-L-lysine (**a**) and fibronectin (**b**). Note the flattened phenotype of fibronectin attached cells in in the presence of serum. Scale bar: 10 μm. Adapted and modified from [9]

adhesion signaling, which allows advanced cell spreading [29]. The cell packing observed on the fibronectin 10 μm squares (Fig. 8c) is limited by adhesion/spreading competition. Upon attachment, the cells readily spread which allows an adherent cell to invade neighboring, free fibronectin squares, as shown in the magnified image in the inset of Fig. 8c, where one cell occupies four neighboring fibronectin squares.

3.3 Neural Stem Cell Patterning on Microspotted Arrays

Firstly, we were interested in selecting the best ECM protein and its optimal concentration in term of cell adhesion and quality of samples during all the workflow of cell culturing procedure (cell seeding, medium changing, fixation) and collecting the data (staining and immunofluorescence). We have tested three (21, 42, and 84 μg/mL) concentrations of proteins: fibronectin, laminin, and collagen IV, which are the ECM proteins found in the neural stem cell niche in the brain. After cell seeding in the 5 × 10⁴/cm² cell density and the 24 h of culture in the medium containing 2% of FBS, the best adhesion and optimal cell growth was observed on fibronectin, even in the lowest protein concentration. The optimal domain stability during the several day culture and sustaining/not promoting of advanced directed neuronal or glial differentiation enables to use the fibronectin as a reference protein for further analysis of influence of signaling proteins on HUCB-NSC fate decisions (Fig. 10a) [25].

The following step was to optimize the spotting protocol for the fabrication of ECM protein microarrays with fibronectin for HUCB-NSC cell attachment in terms of spot size and array layout using fibronectin as protein model [25].

The concentration of 84 μg/mL of fibronectin and the spot size of 4 microspotted drops were chosen for experiments with neural stem cells (Fig. 11).

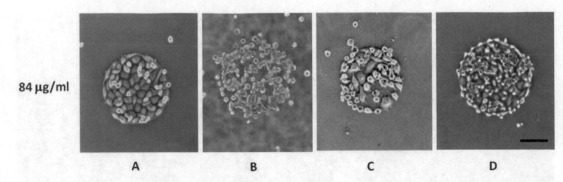

Fig. 10 HUCB-NSCs on fibronectin (**a**), laminin (**b**), collagen V (**c**), and vitronectin (**d**) after 4 days of culture on 3-drop protein spots. The effect of different protein types is visible. Fifty thousands of cells were seeded and maintained in standard culture conditions in presence of 2% serum. The scale bar is 100 μm. Adapted and modified from [25]

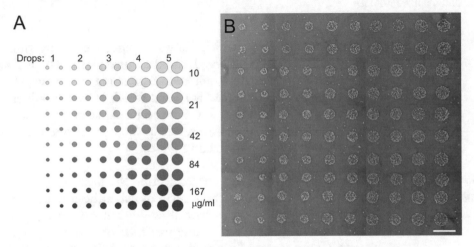

Fig. 11 The spots of fibronectin array were obtained by dispensing different number of drops (from *left*: 1–5 drops, two columns for each drop number) and fibronectin concentrations (from *top*: 10, 21, 42, 84, 167 μg/mL, two rows for each concentration) (**A**). HUCB-NSCs adhesion to the array (10 × 10) of fibronectin spots on PEO-coated slides, after 1 day of cell culture (**B**). Adapted from [8]

3.4 Neural Stem Cell Fate Control on Microprinted Arrays

Microarray layouts of different geometries and biomolecules were prepared to control stem cell adhesion and differentiation. Fate decision of the HUCB-NSC was controlled by the composition and geometry of the active domains. While electrostatic interaction of the cells with the poly-L-lysine kept them non-differentiated, specific integrin receptor -driven interaction with fibronectin, stimulated neuronal differentiation. This was further enhanced by the serum withdraw in the presence of the neuromorphogene—dBcAMP and the specific geometry of the domain—interconnected lines directed neuronal outgrowth.

3.4.1 Proliferation and Migration of NSC on Microprinted Patterns

To estimate the effect of different species of bioprinted domains on proliferation of HUCB-NSC, cells were cultured on the surface patterned with FN or PLL (Fig. 12). Such pattern enables to capture cells on the defined squares of 100 μm side with a 100 μm gap between the squares.

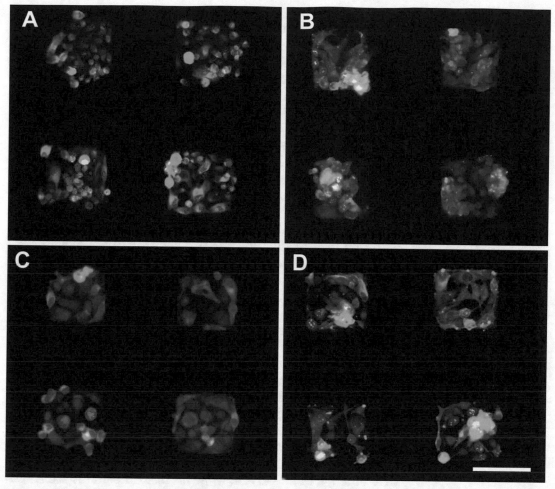

Fig. 12 Immunofluorescence images of HUCB-NSC cells plated on either poly-L-lysine (**a**, **b**) or fibronectin (**c**, **d**) patterns at 2.5 × 10⁴ cells/cm² plating density in 2% of serum (**a**, **c**) and serum free (**b**, **d**) medium at day 4. Cells grown on poly-L-lysine show higher proliferation rate (Ki-67-positive *red dots* in *blue* stained nuclei) than on fibronectin. The β-III tubulin-positive cells are presented in *green*. Cell nuclei are contra-stained with Hoechst (*blue*). Scale bar: 100 μm. Adapted from [9]

The proliferation assay was performed at different seeding density and in the serum free and low serum culture medium [9]. We have measured the percentage of proliferating (Ki67 positive HUCB-NSC) cells per deposited microdomain. We reported that the proliferation rate was higher on poly-L-lysine than on fibronectin printed domains, and poly-L-lysine was able to immobilize more cells on square domains at the same size (Fig. 12). Moreover, the PLL domains, especially in serum free medium, kept the cells at round morphology in a proliferating, non-differentiating state while fibronectin domains of the same size allowed appearance of flattened, strongly attached cells, displaying long protrusions and outgrowth, and revealing lower proliferation rate than on poly-L-lysine, however still at the level even up to 60% of Ki67 positive cells. The ECM protein also promoted neural differentiation of HUCB-NSC cultured on microcontact printed biofunctional arrays [9]. To test whether the deposited biomolecules are able to influence the migration capacity of HUCB-NSC we have performed 24-h observation of HUCB-NSC migrating on micropatterned squares with interconnecting lines printed with either pol-L-lysine or fibronectin. We found that electrostatic type of adhesion is less intensive/strong allowing for cell movement from one domain to another, while ECM protein keeps the cells in the place not allowing for long distance migration [7].

3.4.2 Differentiation of NSC on Microprinted Patterns of Different Size and Biomolecule

Microarray layouts of different geometries (lines and posts) and biomolecules (PLL and FN) were prepared to control stem cell adhesion and differentiation. The layout of one cell resolution posts and lines of PLL and FN revealed that PLL accommodates non-differentiated cells, while FN promotes spreading and neural differentiation. Nestin immunostaining of HUCB-NSC on PLL 10 μm posts (Fig. 13a) and lines (Fig. 13b) confirmed that observation. PLL functionalization was useful for keeping the cells in nestin positive, non-differentiated state (Fig. 13), while the cells grown on fibronectin were negative for nestin immunostaining, indicating that they are at early stages of neuronal or astrocytic differentiation (not shown). Well-defined one-cell resolution patterns of cells with spherical morphology were identified on PLL 10 × 10 μm squares (Fig. 8a). Together with the positive results for nestin staining (Fig. 13), this indicates that PLL one-cell resolution pattern retained the most homogeneous non-differentiated stem cell populations. On PLL patterns of 10 μm wide lines most of the cells were in range of 10–40 μm, while fibronectin patterns accommo-

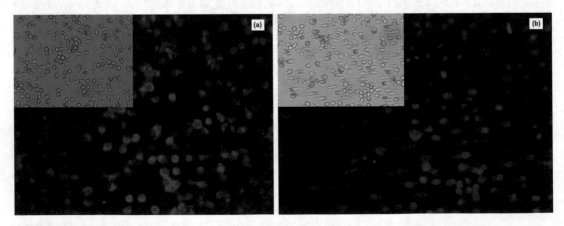

Fig. 13 Fluorescence microscopy images of nestin immunostaining (*red*) of HUCB-NSC after 7 days culturing on PLL *squares* (**a**) and *lines* (**b**). Insets show the corresponding bright field images. Cell nuclei (*blue*) are stained with Hoechst 33342. Adapted and modified from [28]

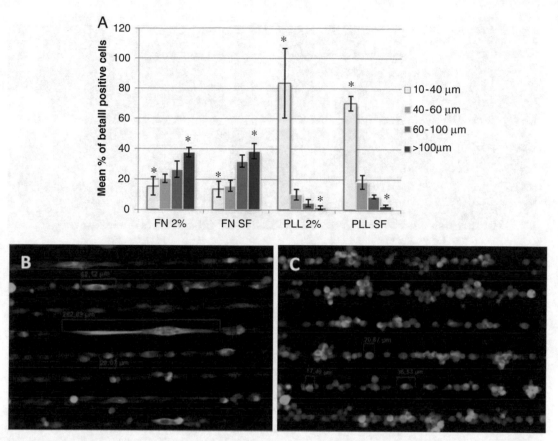

Fig. 14 Distribution of β-tubulin III-positive cells with different cell length parameters, after 24 h of culture (**a**). Immunostaining of β-tubulin III positive cells (*green*) and length measurement of HUCB-NSC cultured on fibronectin (**b**) and poly-L-lysine (**c**) patterns of 10 μm *wide lines*. Cell nuclei are stained with Hoechst (*blue*). Statistically significant difference was observed between the results of the shortest and the longest cells in all experimental groups. *$P < 0.0001$. Adapted and modified from [18]

dated cells longer than 100 μm (Fig. 14a, b). Thus, typical neuronal morphology (longitudinal cells adjusting to the printed lines) was observed only on the fibronectin domains (Fig. 14).

The domains: 120 × 120 μm squares or squares with interconnecting lines immobilized to the surface defined, small population of neural stem cells. We have observed that fibronectin promote cell spreading and neuronal differentiation (presence of neuronal and astrocytic markers—β-tubulin III and S100b, GFAP respectively). Interconnecting 10 μm wide lines enhanced this effect by directing axonal outgrowth (Fig. 15e–h). On the other hand PLL allows for immobilization of non-differentiated, rounded cells (Fig. 15a–d). Addition of dBcAMP further promoted neuronal and astrocytic differentiation in every tested condition.

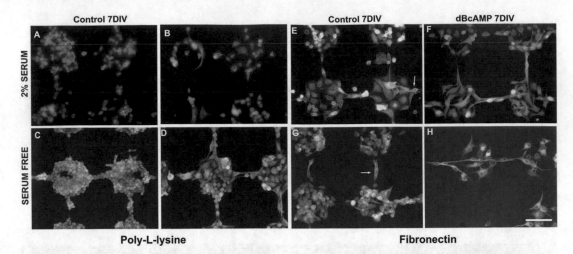

Fig. 15 Immuno-images of HUCB-NSC grown on PLL (**a–d**) and fibronectin (**e–h**) microarrays for 7 days in serum (**a, b, e, f**) and serum free conditions (**c, d, g, h**) differentiating in the presence of dBcAMP (**b, f, d, h**) or growing as a control cultures (**a, c, e, g**). Neuronal marker β-tubulin III is shown *green* for all images. Astrocytic markers are stained *red* for S100β (**a–d**) and GFAP (**e–h**). Cells co-expressing GFAP and β-tubulin III reveal an *orange color* (**e, g**, *arrows*). Cell nuclei are contra-stained with Hoechst (*blue*). The scale bar is 100 μm for all images. Adapted from [26]

3.5 Neural Stem Cell Fate Control on Microspotted Domains

3.5.1 Characterization of Microspotted Domains

Microarray layouts of different content of biomolecules were prepared to control neural stem cell differentiation. As it was mentioned previously, we have chosen the fibronectin as optimal protein and 84 μg/mL as best concentration in terms of cell adhesion and quality of samples during the entire cell culturing procedure (cell seeding, medium changing, fixation) and data collection (staining and immunofluorescence). We were interested whether signaling molecules, such as CNTF, Notch, Jagged, Wnt3a, Shh, Dkk-1, and BMP4, which all play important roles in the development of the nervous system, can determine the HUCB-NSCs fate. Such data represent important information for the possible functionalization of the biomimetic in vitro cues for controlling cellular differentiation [26].

For that purpose, the solution of fibronectin with concentration 84 μg/mL was used as a control spot, and the signaling molecules, such as CNTF, Notch, Jagged, Wnt3a, Shh, Dkk-1 and BMP4 (R&D Systems) were diluted in fibronectin containing spotting buffer and dropped as an array of spots containing different concentrations of each protein representing different signaling molecules (2.5, 5, and 10 μg/mL) on one glass slide.

Verification of Quality of Microspotted Experimental Domains by Ellipsometry

To test the stability of microspotted array during the washing procedure, after preparing the microspotting on the PEO-like coated silicon slide, we have incubated the samples in PBS for 1 h in incubator. After rinsing in water and drying under nitrogen flow the arrays were tested by the ellipsometry. The results are shown as the thickness maps of deposited fibronectin, signaling molecules in

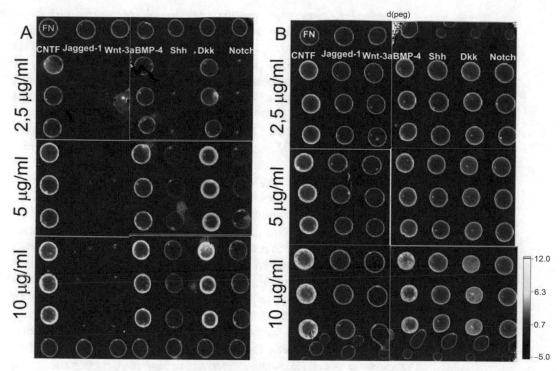

Fig. 16 Thickness map of fibronectin and signaling molecules spotted on plasma-deposited PEO-like film on silicone substrate after rinsing in PBS and measured from ellipsometric Δ and ψ map. The tickness is given in nm with respect to PEO-like deposited on Si master. Signaling molecules spotted alone in printing buffer (**A**), signaling molecules spotted together with fibronectin in printing buffer (**B**)

different concentration and signaling molecules at different concentration mixed with fibronectin solution in printing buffer.

The ellipsometry results showed that after PBS washing in control domains (fibronectin in 84 µg/mL, Fig. 16b, the upper and lower rows) there is a certain amount of the protein that ensures the adhesion of cells to the domains. The control spots containing only the signaling proteins in printing buffer, but lacking the fibronectin (Fig. 16a), were found to contain a certain amount of signaling proteins deposited within the domains (not detected for Jagged-1 and Wnt-3a proteins). However this amount and content was not sufficient for cell adhesion. Fibronectin, which was added to the spotting solution with the SM guarantee the cell resistance and diminish the "ring effect" that is a very common phenomenon in spotting experiments, thus ensuring more uniform protein distribution (Fig. 16b).

Verification of Functionality of Microspotted Experimental Domains by SPR

Several features of bioactive domains have to be considered—the real presence of investigated proteins, e.g., signaling molecules in the domains, the availability of amino acid sequence that is responsible for integrin receptor binding to the extracellular matrix protein and the accessibility of these proteins to the cells' receptors or ligands. For instance, improper protein folding would not allow for the cell responsiveness. The functional analysis—the biological activity was performed using surface plasmon resonance (SPR) imaging in a GenOptics (France) equipment. Monochromatic laser source of $\lambda = 810$ nm measured the changes in the refractive index between gold prism and the dielectric surface). The bioactivity of tested domains was performed on the gold prism surface covered with PEO-like film and deposited with array of spotted fibronectin and fibronectin + SM, using adsorption of antibodies that detect sequence of the deposited biofunctional domains' proteins in respect to the PBS flow, as a reference. To test whether the fibronectin and signaling molecules are within the domains and are biologically active we have used the flow of antibody against the RGD sequence of fibronectin (Ab-RGD), that is necessary for the binding by integrin receptors on the cell surface (Fig. 17) or antibodies specific for individual SM that were spotted with or without fibronectin in printing buffer (Fig. 18). The negative control was performed by flow of bovine serum albumin (BSA). The kinetic curves obtained during measurement by SPR showed adsorption of proteins over time and was detected from each spots/domains that individually react with the antibody (Fig. 17a). The final results expressed as ΔR (%) is the reflectivity difference between measured sample and the baseline (recorded firstly for PBS flow) (Fig. 17b).

We have shown that the amount of the Ab-RGD bounded to the FN domains with signaling molecules is lower than in samples/domains consisting of fibronectin alone (Fig. 17b). This result indicates that the addition of signaling molecules to the printing buffer mixed with fibronectin cause covering/masking of the RGD sequence by the experimental proteins.

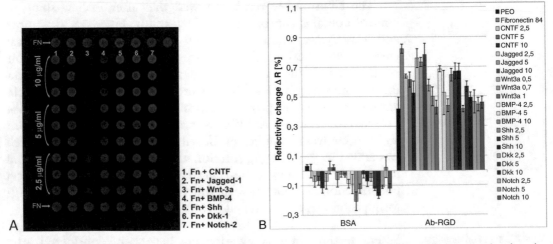

Fig. 17 SPR difference image of the fibronectin (first and last row of the array) and signaling molecules domains after interaction with antibody recognizing the RGD sequence of fibronectin (**a**). The reflectivity difference due to injection of 200 µg/mL BSA, and 10 µg/mL mouse anti-fibronectin RGD-specific antibody (Ab-RGD) measured on spots deposited with different concentration of each signaling molecules (2.5, 5, 10 µg/mL) (**b**)

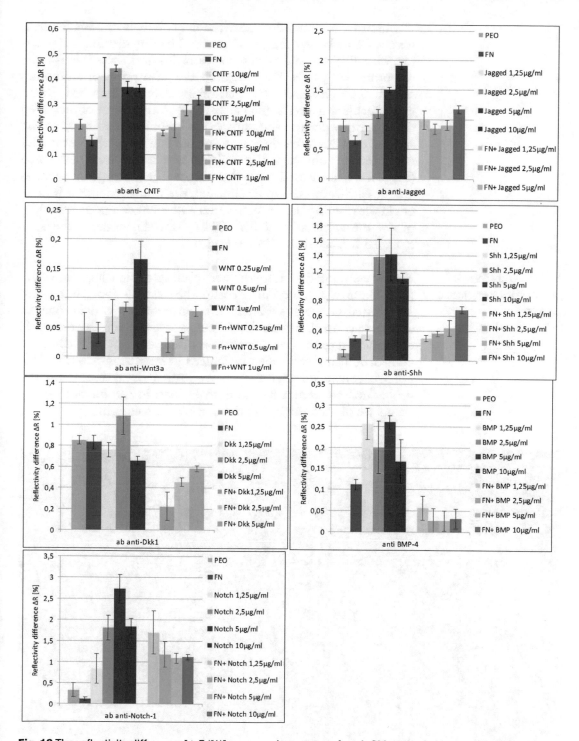

Fig. 18 The reflectivity difference [ΔR (%)] measured on arrays of each SM spotted alone in different concentration and in solution of signaling molecules with fibronectin (FN)

In order to determine the availability of deposited SM for cells, we checked whether the proteins after 2 h of incubation and washing with PBS are still within the domains and accessible for cell receptors or ligands. We measured the adsorption of antibodies specific for every SM on the arrays separate for each SM. These experiments were performed after the BSA treatment of measured domains in order to block unspecific binding of antibody to the spotted proteins, after these washing the flow of specific antibodies were carried out and the reflectivity change was performed (Fig. 18). These experiments showed that the investigated proteins are present within the spots although the fibronectin, that is present in the printing buffer/as a carrier and ECM component slightly masks the selected proteins (Dkk1, BMP4), at least the sites detected by specific to SM antibodies.

Verification of Adhesion of NSC to the Microspotted Domains

Firstly, we wanted to test whether spotting of signaling molecules (SM) alone in printing buffer ensure the cell adhesion. To test this, we spotted arrays of different SM concentration and fibronectin alone as a control. After 5 days of culture we did not observe stable cell adhesion to the domains microspotted only with SM, while the control domains were evenly occupied by neural stem cells (Fig. 19).

We concluded that the carrier protein is necessary for the proper cell adherence to the domains and in the further experiments we diluted the SM in the fibronectin solution.

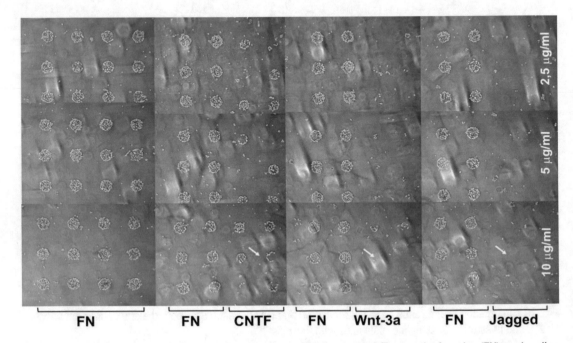

Fig. 19 Optical microscopy of cells seeded on the arrays of microspotted fibronectin domains (FN) or signaling molecules (CNTF, Wnt3a, Jagged) in different protein concentration (2.5; 5 and 10 μg/mL). Note the lack of immobilized HUCB-NSCs to the domains spotted with only the signaling molecules (*arrows*)

3.5.2 Differentiation of NSC on the Biofunctional Domains Microspotted with Signaling Molecules

The structural and functional analysis of biofunctional domains containing signaling molecules were performed for three different concentrations of experimental proteins The presence of signaling molecules was detected even in the domains with the lowest protein concentration (2.5 µg/mL). However, for the investigation of direct of cell differentiation we decided to analyze the highest SM concentration (10 µg/mL) as a variant of the strongest effect.

In order to measure the degree of HUCB-NSC differentiation on the domains we have performed the immunofluorescence of β-tubulin III (as a neuronal) and GFAP (as astrocytic) markers on cells after 4 days of culture. The cells grown on fibronectin spots were used as a reference, since they did not express advanced neuronal and astrocytic markers, remaining non-differentiated. Based on immunocytochemistry analysis we have identified four groups of cells: undifferentiated, that were negative for investigated markers, astrocytes, that expressed GFAP, neurons, that expressed β-tubulin III and cells that co-expressed these markers (transitional phenotype) (Fig. 20). Fibronectin-spotted domains were characterized as reference because of even distribution of certain phenotypes, without statistical significant differences between them. The most potent and statistically significant astrocytic differentiation of HUCB-NSCs was identified on domains comprising: CNTF (33.95% GFAP positive cells), Jagged-1 (31.46%), Notch (33.37%, $p < 0.001$). The lowest percentage of GFAP positive cells was observed on the domains with Wnt-3a (13.81%). The most potent and statistically significant neuronal differentiation, with the highest percentage of β tubulin III positive cells was demonstrated on spots having Shh protein (37.01%), Dkk-1 (35.22%), Wnt-3a (32.63%) and BMP-4 (32.59%), while the lowest level of neuronal differentiation was detected in the presence of CNTF (21.13%). The astro-neuronal (transitional) phenotype of cells was evenly distributed in every investigated variant. Cells that were negative for both markers were detected on all of tested bioactive surfaces; however, we have observed the highest percentage of GFAP/β-tubulin III negative cells on Wnt-3a spots (35.32%). A statistically significant difference in the number of cells expressing astrocytic marker (GFAP) was found in the domains containing CNTF, Notch and Jagged, while the neuronal marker was predominantly observed on the spots plotted with Dkk-1, Wnt3a and BMP-4. Meanwhile, Wnt3a and SHH remained in the nondifferentiating and self-renewing stage. Such data represent important information for the possible functionalization of the biomimetic in vitro cues for controlling cellular differentiation.

4   Conclusions

This study shows that patterning of adhesion molecules by microcontact printing or piezoelectric spotting, allows for attachment and differentiation of NSCs and can be used to direct their fate.

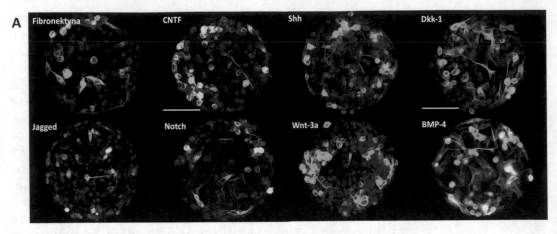

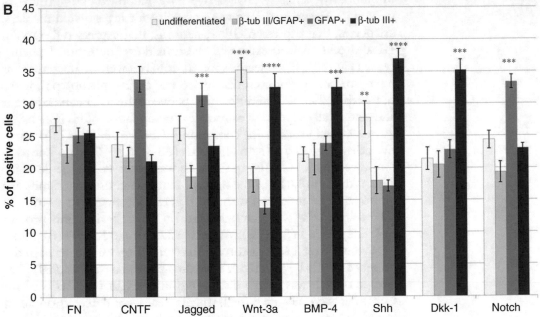

Fig. 20 (**a**) Immunofluorescence imaging of HUCB-NSCs cultured on domains microspotted with different signaling molecules, and stained for β -tubulin III (*green*) and GFAP (*red*) antibodies. Cell nuclei are contra stained with Hoechst 33258 (*blue*). Scale bar = 100 μm. (**b**) HUCB-NSCs differentiation on tested arrays. The phenotype distribution is present as mean % (±SEM) of positive cells for investigated markers. The statistical significant difference is measured in relation to % of positive cells cultured on fibronectin, calculated from five independent experiments (One-Way ANOVA with post hoc Dunnett test, ***$p < 0.001$, **$p < 0.01$)

Biofunctional domains with the properties for nonspecific, electrostatic cell membrane–surface interaction (PLL based) versus receptor mediated specific interaction (ECM based) were designed and fabricated with different content of biomolecules and geometry. The ellipsometry and SPR measurements revealed that deposited molecules in bioactive domains are present and functional. They allow for the immobilization of the non-differentiated and highly proliferating population of neural stem/progenitor cells to the biofunctionalized surface or directional differentiation into neuronal lineage and guiding of the axonal outgrowth. The single cell patterning system (posts and lines) confirmed the influence of pattern geometry on the mode of neural stem cell differentiation. Functional spotted domains were further engineered to create "smart" microenvironment by immobilizing to the surface small signaling molecules together with ECM proteins. Stimulation of selected intracellular pathways by signaling molecules: Wnt, Shh, CNTF, Jagged, Notch, or BMP-4 type resulted in different fate decisions of HUCB-NSC. Wnt-3 and Shh support proliferation and self-renewal, but also neuronal commitment of HUCB-NSCs. Jagged as the ligand and stimulator of Notch signaling pathway and CNTF acting through JAK/STAT directed HUCB-NSC into astrocytic lineage, while Dkk1 and BMP-4 have supported neuronal differentiation. This type of bioengineered cell growth platforms allow for screening the mechanisms governing neural stem cell fate decisions and adverse reactions upon environmental stimuli.

Acknowledgments

We are grateful to Dr Laura Ceriotti for the participation in the microspotting project. We also thank Sabrina Gioria, Patricia Lisboa, G. Ceccone, and Hubert Rauscher for their scientific and technical assistance. This work was supported by statutory founds to MMRC and by Polish Ministry of Scientific Research and Higher Education grant No: 2211/B/P01/2010/38 and European Commission Joint Research Centre Actions "NanoBiotechnology for Health" and "Validation for Consumer Products."

References

1. Lutolf MP, Gilbert PM, Blau HM (2009) Designing materials to direct stem-cell fate. Nature 462(7272):433–441

2. Stabenfeldt SE, Munglani G, García AJ, LaPlaca MC (2010) Biomimetic microenvironment modulates neural stem cell survival, migration, and differentiation. Tissue Eng Part A 16(12):3747–3758

3. Riquelme PA, Drapeau E, Doetsch F (2008) Brain micro-ecologies: neural stem cell niches in the adult mammalian brain. Philos Trans R Soc Lond B Biol Sci 363(1489):123–137

4. Lee MR, Kwon KW, Jung H, Kim HN, Suh KY, Kim K, Kim KS (2010) Direct differentiation of human embryonic stem cells into selective neurons on nanoscale ridge/groove pattern arrays. Biomaterials 31(15):4360–4366

5. Vecino E, Heller JP, Veiga-Crespo P, Martin KR, Fawcett JW (2015) Influence of extracellular matrix components on the expression of integrins and regeneration of adult retinal ganglion cells. PLoS One 10(5):e0125250

6. Kazanis I, Lathia JD, Vadakkan TJ, Raborn E, Wan R, Mughal MR, Eckley DM, Sasaki T, Patton B, Mattson MP, Hirschi KK, Dickinson ME, ffrench-Constant C (2010) Quiescence and activation of stem and precursor cell populations in the subependymal zone of the mammalian brain are associated with distinct cellular and extracellular matrix signals. J Neurosci 21:9771–9781

7. Ruiz A, Buzanska L, Gilliland D, Rauscher H, Sirghi L, Sobanski T, Zychowicz M, Ceriotti L, Bretagnol F, Coecke S, Colpo P, Rossi F (2008) Micro-stamped surfaces for the patterned growth of neural stem cells. Biomaterials 29(36):4766–4774

8. Bużańska L, Zychowicz M, Ruiz A, Ceriotti L, Coecke S, Rausher H, Sobanski T, Wheland M,

Domanska-Janik K, Colpo P, Rossi F (2010) Neural stem cells from human cord blood on bioengineered surfaces—novel approach to multiparameter bio-tests. Toxicology 270:35–42

9. Zychowicz M, Mehn D, Ana Ruiz A, Colpo P, Francois Rossi F, Frontczak-Baniewicz M, Domanska-Janik K, Buzanska L (2011) Proliferation capacity of cord blood derived neural stem cell line on different micro-scale biofunctional domains. Acta Neurobiol Exp (Wars) 71(1):12–23

10. Béduer A, Vieu C, Arnauduc F, Sol JC, Loubinoux I, Vaysse L (2012) Engineering of adult human neural stem cells differentiation through surface micropatterning. Biomaterials 33(2):504–514

11. Joo S, Kim JY, Lee E, Hong N, Sun W, Nam Y (2015) Effects of ECM protein micropatterns on the migration and differentiation of adult neural stem cells. Sci Rep 5:13043

12. Saha K, Pollock JF, Schaffer DV, Healy KE (2007) Designing synthetic materials to control stem cell phenotype. Curr Opin Chem Biol 11(4):381–387

13. Bakhru S, Nain AS, Highley C, Wang J, Campbell P, Amon C, Zappe S (2011) Direct and cell signaling-based, geometry-induced neuronal differentiation of neural stem cells. Integr Biol (Camb) 3(12):1207–1214

14. Dertinger SK, Jiang X, Li Z, Murthy VN, Whitesides GM (2002) Gradients of substrate-bound laminin orient axonal specification of neurons. Proc Natl Acad Sci USA 99(20):12542–12547

15. Millet LJ, Stewart ME, Nuzzo RG, Gillette MU (2010) Guiding neuron development with planar surface gradients of substrate cues deposited using microfluidic devices. Lab Chip 10(12):1525–1535

16. Dalby MJ, Andar A, Nag A, Affrossman S, Tare R, McFarlane S, Oreffo RO (2008) Genomic expression of mesenchymal stem cells to altered nanoscale topographies. J R Soc Interface 5(26):1055–1065

17. Kim WR, Jang MJ, Joo S, Sun W, Nam Y (2014) Surface-printed microdot array chips for the quantification of axonal collateral branching of a single neuron in vitro. Lab Chip 14(4):799–805

18. Zychowicz M, Mehn D, Ruiz A, Frontczak-Baniewicz M, Rossi F, Buzanska L (2012) Patterning of human cord blood-derived stem cells on single cell posts and lines: implications for neural commitment. Acta Neurobiol Exp (Wars) 72(4):325–336

19. Soen Y, Mori A, Palmer TD, Brown PO (2006) Exploring the regulation of human neural precursor cell differentiation using arrays of signaling microenvironments. Mol Syst Biol 2:37

20. Bużańska L, Jurga M, Stachowiak EK, Stachowiak MK, Domanska-Janik K (2006) Neural stem-like cell line derived from a non-hematopoietic population of human umbilical cord blood. Stem Cells Dev 15:391–406

21. Ruiz A, Zychowicz M, Ceriotti L, Mehn D, Sirghi L, Rauscher H, Mannelli I, Colpo P, Buzanska L, Rossi F (2013) Microcontact printing and microspotting as methods for direct protein patterning on plasma deposited polyethylene oxide: application to stem cell patterning. Biomed Microdevices 15(3):495–507

22. Brétagnol F, Lejeune M, Papadopoulou-Bouraoui A, Hasiwa M, Rauscher H, Ceccone G, Colpo P, Rossi F (2006) Fouling and non-fouling surfaces produced by plasma polymerization of ethylene oxide monomer. Acta Biomater 2(2):165–172

23. Bretagnol F, Ceriotti L, Lejeune M, Papadopoulou A, Hasiwa M, Gilliland D, Ceccone G, Colpo P, Rossi F (2006) Functional micropatterned surfaces by combination of plasma polymerization and lift-off processes. Plasma Process Polym 3(1):30–38

24. Ruiz A, Ceriotti L, Buzanska L, Hasiwa M, Bretagnol F, Ceccone G, Gilliland D, Rauscher H, Coecke S, Colpo P, Rossi F (2007) Controlled micropatterning of biomolecules for cell culturing. Microelectron Eng 84:1733–1736

25. Ceriotti L, Buzanska L, Rausche rH, Mannelli I, Sirghi L, Gilliland D, Hasiwa M, Bretagnol F, Zychowicz M, Ruiz A, Bremer S, Coecke S, Colpo P, Rossi F (2009) Fabrication and characterization of protein arrays for stem cell patterning. Soft Matter 5:1406–1416

26. Buzanska L, Ruiz A, Zychowicz M, Rausher H, Ceriotti L, Rossi F, Colpo P, Domanska-Janik K, Coecke S (2009) Patterned growth and differentiation of human cord blood-derived neural stem cells on bio-functionalized surfaces. Acta Neurobiol Exp (Wars) 69:1–14

27. Guemouri L, Ogier J, Zekhnini Z, Ramsden JJ (2000) The architecture of fibronectin at surfaces. J Chem Phys 113:8183

28. Ruiz A, Zychowicz M, Buzanska L, Mehn D, Mills CA, Martinez E, Coecke S, Samitier J, Colpo P, Rossi F (2009) Single stem cell positioning on polylysine and fibronectin microarrays. Micro Nanosyst 1:50–56

29. Margadant C, van Opstal A, Boonstra J (2007) Focal adhesion signaling and actin stress fibers are dispensable for progression through the ongoing cell cycle. J Cell Sci 120:66–76

Chapter 3

Dopaminergic and GABAergic Neuron In Vitro Differentiation from Embryonic Stem Cells

Talita Glaser, Juliana Corrêa-Velloso, Ágatha Oliveira-Giacomelli, Yang D. Teng, and Henning Ulrich

Abstract

Embryonic stem cells (ESCs) are an important resource for translational medicine, cell therapy applications and provide an established model for nontherapeutic applications such as in vitro pharmacological and toxicological screening of neuroactive compounds as well as in vitro modeling of neurodegenerative and neurodevelopmental diseases. Relevant progress in this field led to the establishment of efficient stepwise differentiation protocols for the generation of neuronal progenitor cells (NPCs) that can undergo differentiation into specific neuronal subtypes that simulate the morphological and physiological characteristics of their in vivo counterparts.

In order to elucidate the mechanisms of disease establishment or cell therapy technology in some neurodegenerative diseases such as Parkinson's and Huntington's disease, protocols based on monolayer induction or embryoid body (EB) formation, followed by neural selection and specific phenotype induction, were developed to obtain dopaminergic or GABAergic neurons, respectively.

In view of that, in this chapter we describe two protocols based on EB formation and achievement of NPCs to obtain dopaminergic or GABAergic neurons from mouse ESCs.

Key words Neuron, Stem cell, Neural precursor cell, Phenotype

1 Introduction

Pluripotent stem cells [embryonic stem cells (ESCs) and induced pluripotent stem cells (iPSCs)] can differentiate into all cell types present in the adult organism and can indefinitely self-renew under in vitro conditions. These properties make them a relevant resource, both for investigating early developmental processes and for assessing their therapeutic potential in several models of degenerative diseases. Since successful mouse and human ESC isolation [1, 2], relevant progress has been made in this field, principally based on the establishment of efficient stepwise differentiation protocols for the generation of neuronal progenitors (NPCs) from pluripotent stem cells. These progenitors can undergo differentiation into

Amit K. Srivastava et al. (eds.), *Stem Cell Technologies in Neuroscience*, Neuromethods, vol. 126,
DOI 10.1007/978-1-4939-7024-7_3, © Springer Science+Business Media LLC 2017

specific neuronal subtypes reflecting morphological and physiological characteristics of their in vivo counterparts.

Neural differentiation of ESCs is usually achieved by induction of ectoderm fate followed by the enrichment of neuronal progenitors using a variety of peptide factors [3]. Along neural fate specification, ESCs lose expression of pluripotent markers, such as octamer-binding transcription factor 4 (Oct4) and Nanog homeobox, while gaining neuronal-specific phenotype markers, like: Nestin for neural precursor cells, neuron-specific class III beta-tubulin (TUJ1) for young neurons, dopamine- and cyclic AMP-regulated phosphoprotein (DARPP32) and glutamate decarboxylase (GAD65 and 67) for GABAergic medium spiny neurons [4], and tyrosine hydroxylase (TH) for dopaminergic neurons [5]. Expression levels of these markers can be determined through several molecular techniques in order to both track maturation progression and assess the differentiation into neuronal subtypes.

The control of neuronal differentiation is a reliable source for translational medicine and cell therapy applications, such as the evaluation of regenerative function and integration of grafted neurons in animal models of neurodegenerative disease such as Parkinson's and Huntington's disease [4, 6].

Differentiated stem cells are also an established model for non-therapeutic applications such as in vitro pharmacological and toxicological screening of neuroactive compounds [7, 8], as well as in vitro modeling of neurodegenerative and neurodevelopmental diseases [9].

In order to establish in vitro stem cell differentiation models that resemble the neurodevelopment or that generate high purity of specific neuron phenotype, many research groups focused on Parkinson's and Huntington's disease have published different protocols to obtain dopaminergic and GABAergic neurons, respectively. Most of these protocols are based on embryoid body (EB) formation, which consists of a sphere of cells similar to a blastocyst at a late stage [10]. These cells are able to originate tissues of the three germ layers (mesoderm, ectoderm, and endoderm), and the addition of all-trans retinoic acid at specific concentrations directs the differentiation preferably to neuroectodermal lineages [11]. There are also protocols referring to selection and expansion of NPCs, due to their ability to survive at serum free conditions, differently from other cells types [12]. Furthermore, there are some methods based on the differentiation induction of a monolayer ESCs culture without EB formation stage [13].

In this chapter, we describe two protocols based on EB formation and sequential selection of NPCs to obtain dopaminergic or GABAergic neurons from differentiating mouse ESCs. The protocols follow the same initial steps, but slightly differ on the culture medium composition in the late two stages, turning to be a helpful tool to direct cell differentiation depending on the research hypothesis.

2   Materials

All reagents and materials must be previously sterilized. Prepared reagents should be stored at 4 °C unless otherwise stated.

Phosphate-buffered saline (PBS) 1x: phosphate buffer saline (137 mM NaCl, 10 mM Na_2HPO_4, 2.7 mM KCl, 1.8 mM KH_2PO_4). Dissolve the reagents listed above in water. Adjust the pH to 7.4 with HCl.

Trypsin solution: 0.25% trypsin, 1 mM EDTA (ethylenediamine tetraacetic acid), pH 7.4. Filter through a 0.22 μm pore membrane.

StemPro® Accutase® Cell Dissociation Reagent from Thermo Fisher Scientific.

Embryoid body differentiation (EB) medium: Please check Table 1.

Neural precursor cell selection medium (NPC-S) medium: Please check Table 1.

Dopaminergic precursor cell expansion medium (DPC-E) medium: Please check Table 1.

Dopaminergic neuron differentiation medium (DN-D) medium: Please check Table 1.

GABAergic precursor cell expansion medium (GPC-E) medium: Please check Table 1.

GABAergic neuron differentiation medium (GN-D) medium: Please check Table 1.

3   Methods

3.1   Cell Manipulation

Step I—Maintenance of feeder free ESCs

There are many different mouse ESC lineages commercially available. In this chapter, we use the ESC E14tg2a cell line: 129/Ola-derived HPRT-negative ESC. The protocol of this cell lineage maintenance is well described in other book chapters [14].

Step II—Embryoid bodies' formation

1. Rinse ESCs with PBS1× without calcium and magnesium. (Note 1).

2. Detach the cells from the dish using trypsin solution for 2–5 min at 37 °C.

3. Resuspend ~5 × 10⁵ ESCs in 5 mL EB differentiation medium and seed cells onto low adherent T25mm tissue culture flask for induction of EB formation (Fig. 1) (Note 2).

4. After 2 days of floating culture, sediment cells in a 15 mL conical tube, replace EB differentiation medium.

Table 1
Instructions for medium preparation

Culture medium	Components	Amount/final concentration	Tips
Embryoid body medium (EB) (Step II)	Dulbecco's modified Eagle's medium (DMEM)—high glucose	quantum sufficit (q.s.)	– pH 7.4; – Filter through 0.22 μm pore size membranes. – There is a substitute stabilized form of l-glutamine available for purchase.
	ES cell qualified fetal bovine serum (ES FBS)	20% V/V	
	MEM nonessential amino acids (NEAA)	1% V/V	
	Sodium pyruvate	2 mM	
	l-glutamine	2 mM	
	2-mercaptoethanol	0.55 mM	
Neural precursor cell selection medium (NPC-S) (Step III)	DMEM F-12, no HEPES	q.s.	– pH 7.4; – Prepare fresh as needed.
	Insulin	5 μg/mL	
	Apo-transferrin	50 μg/mL	
	Sodium selenite	30 nM	
	Fibronectin	5 μg/mL	
	l-glutamine	2 mM	
Dopaminergic precursor cell expansion medium (DPC-E) (Step IVa)	DMEM F-12, no HEPES	q.s.	– pH 7.2 ± 0.2; – Ascorbic acid is a light-sensitive compound; – Prepare fresh as needed; – Aliquot growth factor solutions to avoid thawing and refreezing; – Commercial mixed supplements available.
	Glucose	1.55 mg/L	
	Ascorbic acid	200 μM	
	Insulin	25 μg/mL	
	Apo-transferrin	50 μg/mL	
	Sodium selenite	30 nM	
	Progesterone	20 nM	
	Putrescine	100 μM	
	l-glutamine	2 mM	
	Basic fibroblast growth factor (bFGF)	10 ng/mL	
	Fibroblast growth factor 8b (FGF-8b)	10 ng/mL	
	Sonic Hedgehog N-Terminus (Shh-N)	10 ng/mL	

(continued)

Table 1
(continued)

Culture medium	Components	Amount/final concentration	Tips
Dopaminergic neuron differentiation medium (DN-D) (Step Va)	DMEM F-12, no HEPES	q.s.	– pH 7.4; – Ascorbic acid is a light-sensitive compound; – Commercial supplements available.
	Glucose	1.55 mg/mL	
	Ascorbic acid	200 μM	
	Insulin	25 μg/mL	
	Apo-transferrin	50 μg/mL	
	Sodium selenite	30 nM	
	Progesterone	20 nM	
	Putrescine	100 μM	
	l-glutamine	2 mM	
Gabaergic precursor cell expansion medium (GPC-E) (Step IVa)	DMEM F-12	q.s.	– pH 7.4; – Prepare fresh as needed; – Aliquot growth factor solutions to avoid thawing and refreezing; – Commercial supplements available.
	Insulin	25 μg/mL	
	Apo-transferrin	50 μg/mL	
	Sodium selenite	30 nM	
	Progesterone	20 nM	
	Putrescine	100 μM	
	l-glutamine	2 mM	
	Laminin	1 μg/mL	
	bFGF	10 ng/mL	
Gabaergic neuron differentiation medium (GN-D) (Step Vb)	DMEM F-12	q.s.	– pH 7.4; – Commercial supplements available.
	Insulin	25 μg/mL	
	Apo-transferrin	50 μg/mL	
	Sodium selenite	30 nM	
	Progesterone	20 nM	
	Putrescine	100 μM	
	l-glutamine	2 mM	
	Laminin	1 μg/mL	

Step III—Enrichment of Nestin-positive cells

1. Following two further days, replace EB differentiation medium and seed EBs into laminin pre-coated dishes. Seed 20 EBs/well of a six well plate.

2. After 24 h, replace the EB medium by Neural precursor cell Selection medium (NPC-S), 2 mL/well;

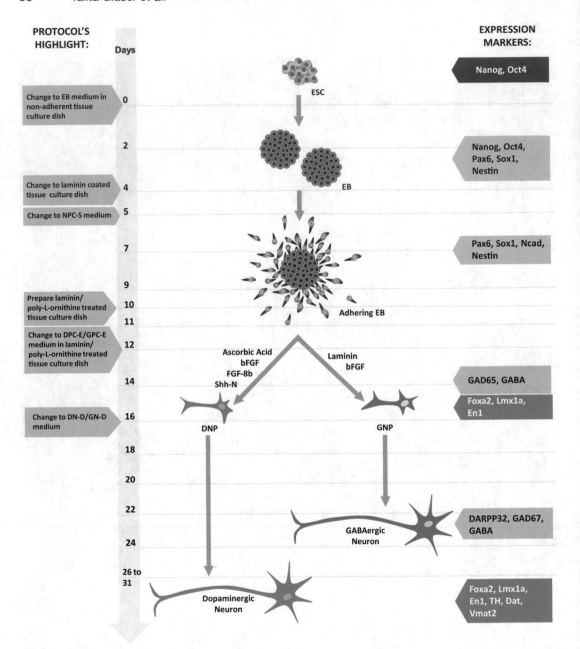

Fig. 1 Outline of differentiation of mouse ESCs into dopaminergic or *GABAergic neurons*. ESCs can differentiate into dopaminergic or GABAergic neurons after the formation of EBs and sequential selection and induction of NPCs. At the *left*, there are some highlights of the protocol, while at the right side of the figure marker proteins typically expressed at respective stages of differentiation are indicated

3. Change medium daily for 6–8 days. A monolayer will grow from the attached EBs.

Step IV—Cell fate commitment to specific neuron phenotypes

From this step on, cell fate direction into dopaminergic or GABAergic neurons can be initiated. For differentiation into dopaminergic neurons, as first described by Okabe et al. [12] and modified by Lee et al. [15], please follow steps IVa and Va. To obtain GABAergic neurons, as described by Shin et al. [4], please follow steps IVb and Vb.

Differentiation Process of Dopaminergic Neurons:
Step IVa—Expansion of Dopaminergic Precursor Cells (DPCs)

1. After selection of NPCs for 6–8 days, rinse cells with PBS without calcium and magnesium.

2. Detach cells from the dish using StemPro® Accutase® Cell Dissociation Reagent for 2–5 min at 37 °C.

3. Add 5 mL medium to cells, transfer and sediment cells for 5 min in a 15 mL conical tube.

4. Collect and transfer the supernatant containing the NPCs to another 15 mL conical tube and centrifuge for 5 min at $220 \times g$.

5. Discard supernatant and resuspend cells in DPC-E medium.

6. Count cells and seed $3–5 \times 10^5$ cells/well, of a 24 well plate, on poly-l-ornithine and laminin pre-coated dishes (Sect. 3.2) in DPC-E medium.

7. Culture for 4–6 days. Change medium daily.

8. Cell confluence should reach close to 100%.

Step Va. Differentiation into Dopaminergic Neurons (DNs)

1. For induction of dopaminergic precursor cell differentiation into dopaminergic neurons, replace the medium by DN-D medium.

2. Culture cells for 10–15 days.

3. Check for anti-tyrosine hydroxylase antibody staining.

Differentiation Process of GABAergic Neurons:
Step IVb—Expansion of GABAergic Precursor Cells (GPCs)

1. After selection of NPCs for 8 days, rinse cells with PBS without calcium and magnesium;

2. Detach the cells from the dish using StemPro® Accutase® Cell Dissociation Reagent for 2–5 min at 37 °C.

3. Add 5 mL medium to cells, transfer and sediment cells in a 15 mL conical tube for 5 min.

4. Collect and transfer the supernatant containing the NPCs to another 15 mL conical tube and centrifuge at $220 \times g$ for 5 min.

5. Discard supernatant and resuspend cells in GPC-E medium.

6. Count cells and seed $3–5 \times 10^5$ cells/well, of a 24-well plate, on poly-l-ornithine and laminin pre-coated dishes in GPC-E medium.

7. Culture for 6 days. Change medium daily.

8. Cell confluence should reach close to 100%.

Step Vb—Differentiation to GABAergic neurons (GNs)

1. For differentiation of GPCs into GNs, replace the medium by GN-D medium.

2. Culture cells for 6 days.

3. Check for the expression of specific markers (Fig. 1).

3.2 Dish Treatment   *Poly -1 - ornithine and Laminin coating*

1. Add a volume 15 µg/mL poly-l-ornithine that completely covers the well of a six wells cell culture plate.

2. Incubate overnight at 37 °C. Dish must not dry.

3. Discard the poly-l-ornithine solution and wash three times with PBS.

4. Add PBS and incubate overnight at 37 °C.

5. Discard PBS.

6. Add 50 µg/mL laminin to the dish and incubate at 37 °C for at least 30 min.

7. Wash 3× with PBS.

8. Laminin can be reused many times.

4 Notes

1. E14tg2a cells are 129/Ola-derived HPRT-negative ESCs; in view of that, HPRT-minigene can be used as an additional option of selectable marker by combination with HAT selection.

2. EBs are already observed 2 days after seeding.

Acknowledgments

This work was supported by the research grants awarded by the São Paulo Research Foundation (FAPESP Proj. No. 2012/50880-4) and the National Council for Scientific and Technological Development (CNPq Proj. No. 467465/2014-2) and by the Provost's Office for Research of the University of São Paulo, Grant number: 2011.1.9333.1.3 (NAPNA-USP), Brazil. Work at Teng Laboratories was supported by CDMRP-DoD, CASIS-NASA, VAR&D, and The Gordon Project of Harvard Medical School, USA.

References

1. Martin GR (1981) Isolation of a pluripotent cell line from early mouse embryos cultured in medium conditioned by teratocarcinoma stem cells. Proc Natl Acad Sci USA 78:7634–7638

2. Thomson JA, Itskovitz-Eldor J, Shapiro SS et al (1998) Embryonic stem cell lines derived from human blastocysts. Science 282:1145–1147

3. Smith AG (2001) Embryo-derived stem cells: of mice and men. Annu Rev Cell Dev Biol 17:435–462

4. Shin E, Palmer MJ, Li M et al (2012) GABAergic neurons from mouse embryonic stem cells possess functional properties of striatal neurons in vitro, and develop into striatal neurons in vivo in a mouse model of Huntington's disease. Stem Cell Rev 8:513–531

5. Salti A, Nat R, Neto S et al (2013) Expression of early developmental markers predicts the efficiency of embryonic stem cell differentiation into midbrain dopaminergic neurons. Stem Cells Dev 22:397–411

6. Kang X, Xu H, Teng S et al (2014) Dopamine release from transplanted neural stem cells in Parkinsonian rat striatum in vivo. Proc Natl Acad Sci USA 111:15804–15809

7. Sison-Young RL, Kia R, Heslop J et al (2012) Human pluripotent stem cells for modeling toxicity. Adv Pharmacol 63:207–256

8. Nantasanti S, de Bruin A, Rothuizen J et al (2016) Concise review: organoids are a powerful tool for the study of liver disease and personalized treatment design in humans and animals. Stem Cells Transl Med 5(3):325–330

9. Yuan SH, Shaner M (2013) Bioengineered stem cells in neural development and neurodegeneration research. Ageing Res Rev 12:739–748

10. Itskovitz-Eldor J, Schuldiner M, Karsenti D et al (2000) Differentiation of human embryonic stem cells into embryoid bodies compromising the three embryonic germ layers. Mol Med 6:88–95

11. Okada Y, Shimazaki T, Sobue G et al (2004) Retinoic-acid-concentration-dependent acquisition of neural cell identity during in vitro differentiation of mouse embryonic stem cells. Dev Biol 275:124–142

12. Okabe S, Forsberg-Nilsson K, Spiro AC et al (1996) Development of neuronal precursor cells and functional postmitotic neurons from embryonic stem cells in vitro. Mech Dev 59:89–102

13. Ying QL, Stavridis M, Griffiths D et al (2003) Conversion of embryonic stem cells into neuroectodermal precursors in adherent monoculture. Nat Biotechnol 21:183–186

14. Glaser T, Castillo AR, Oliveira A et al (2015) Intracellular calcium measurements for functional characterization of neuronal phenotypes. Methods Mol Biol 1341:245–255

15. Lee SH, Lumelsky N, Studer L et al (2000) Efficient generation of midbrain and hindbrain neurons from mouse embryonic stem cells. Nat Biotechnol 18:675–679

Chapter 4

Transfection of Cultured Primary Neurons

Annalisa Rossi, Ralf Dahm, and Paolo Macchi

Abstract

The efficient delivery of genes into neurons is a crucial tool for the study of neuronal cell biology as well as for the development of novel therapeutic approaches, such as gene therapy. Over the past years, numerous techniques have been established to deliver genes into cells. These methods can be broadly classified into two main groups: viral based and nonviral methods. The viral methods use viral particles such as adenoviruses, adeno-associated, lentiviruses, and herpes simplex viruses. Nonviral methods can be subdivided into physical and chemical methods. While the first one includes techniques such as electroporation, magnetofection, microinjection, and biolistics, the latter comprises lipofection and calcium phosphate/ DNA co-precipitation. Each one of these methods has its own advantages and drawbacks and the choice of a particular method depends on the experimental setting and objective. In this chapter, we summarize the key advantages and disadvantages of various techniques for the gene delivery in primary neurons and neuronal stem cells.

Key words Primary neurons, Neuronal progenitor cells, Stem cells, Gene delivery, Transfection, Viral methods, Nonviral methods

1 Introduction

Primary neuronal cultures are commonly used to understand mechanisms of neural function, dysfunction, and degeneration. In particular, primary neuronal cultures are suitable to carry out in-depth studies of molecular events, biochemical pathways or neuronal excitability, which are difficult to perform with in vivo models [1, 2]. Moreover, primary cultures allow easy access to single neurons for high-resolution microscopy analyses, genetic manipulations, electrophysiological recordings, and recently also genomic/ transcriptomic analyses [3–10]. Experiments performed on primary neuronal cultures are important for numerous areas of neurobiology research. For instance, neuronal cell cultures are a useful tool to study neurotoxic effects of drugs [11–13], neuronal plasticity [14–16] as well as the molecular bases of neuronal diseases [17–20].

Amit K. Srivastava et al. (eds.), *Stem Cell Technologies in Neuroscience*, Neuromethods, vol. 126,
DOI 10.1007/978-1-4939-7024-7_4, © Springer Science+Business Media LLC 2017

Although primary neuronal cultures have crucial advantages as regards the controllability, reproducibility, and the possibility to observe, manipulate, and molecularly analyze single cells, they have some limitations [21]. When neurons are dissected from animals and single cell suspensions are prepared, the connections previously established between the cells are lost and new ones are formed in vitro, which mirror the in vivo situation only to some extent. Moreover, when cultured outside their "natural" environment, neuronal cells are influenced in non-physiological ways and their responses will—at least to some extent—differ from the in vivo condition [22]. Another clear disadvantage lies in the fact that neuronal cultures are limited to studying phenomena restricted to single cells or small groups of cells. Therefore, results obtained from neuronal cultures cannot easily be extended to higher levels of function of nervous systems, especially not to complex processes such as cognition and consciousness.

However, despite these drawbacks of working with cells in culture, the delivery of synthetic nucleic acids (e.g., recombinant DNA, RNA and oligonucleotides) into cultured primary neurons and neural stem cells (NSCs), and thus the controlled manipulation of gene expression or protein function in these cells, is a key approach to understand the function of genes involved in neural development, plasticity, and physiology, as well as to elucidate pathological mechanisms and to develop therapeutic applications [23–29]. Neuronal gene transfer can be used, for example, to express epitope-tagged or fluorescently tagged proteins or RNAs, including wild-type or mutated forms, to study their localization, movement, and interactions inside living cells and/or their effects on cell functions. In addition, they can serve as bait for immunoprecipitation or pull-down experiments. Furthermore, gene transfer into neurons has also been used to knockdown specific genes of interest using antisense oligonucleotides or RNA interference (RNAi) approaches. Transfected neurons can then be analyzed through biochemical, imaging, electrophysiological or single cell -omics approaches to verify the molecular and functional consequences of these manipulations on individual neurons. Moreover, NSCs can be stably engineered to express specific genes in a spatio-temporally controlled way. By transplanting genetically modified NSCs into the CNS, it is possible to investigate their behavior, for example their migration in response to morphogen gradients or cytokine signals in injury or disease conditions [30, 31].

The existing gene delivery methods for primary post-mitotic neurons and NSCs have several limitations due to intrinsic characteristics of these cells. The post-mitotic neurons, indeed, are very sensitive to microenvironmental changes. In addition, the foreign nucleic acid cannot enter in the nucleus during the nuclear envelope breakdown, occurring only in mitotic cells. The transfection efficiencies of NCSs are also very low because the transfection procedures are often toxic to these cells [32]. Moreover, these techniques can easily interfere with the NSCs proliferation and differentiation.

Ideally, a transfection method should allow an easy and safe cellular internalization of nucleic acids of various sizes, with high efficiency and minimal cellular toxicity [33]. For microscopy or single-cell transcriptomic analyses of individual cells, a low to medium transfection rate is sufficient. By contrast, for quantitative biochemical analyses, such as immunoprecipitation experiments, high transfection efficiencies are necessary. A potential therapeutic use of NSCs cells requires a controlled and stable expression of genes of interest in order not to be toxic for the implanted organism.

Current gene transfer methods are commonly divided into two major classes: viral- and nonviral-dependent gene transfer. Viral vectors commonly used to transduce primary neurons or NSCs can be classified into four major categories, each one of these with different properties and abilities to either integrate their genome into that of the host cell, such as lentivirus (LV), or persist in the nucleus in an episomal form, such as adenovirus (AdV), adeno-associated virus (AAV), and herpes simplex 1 virus (HSV-1). Nonviral gene transfer methods are classified into two main groups: physical and chemical. The physical methods include electroporation, magnetofection, ballistic gene transfer, and microinjection. The major chemical methods are lipofection and calcium phosphate transfection.

This chapter presents an overview of the most commonly used transfection methods for neuronal cells, provides typical examples of applications and compares the methods discussed regarding their suitability for gene transfer into post-mitotic neurons and NSCs.

2 Viral Gene Transfer

Virus-mediated gene transfer methods are becoming fundamental tools for the delivery of nucleic acids into primary cell cultures and in vivo models. This approach is highly efficient and allows sustainable transgene expression both in vitro and in vivo thanks to the viruses' natural ability to infect cells and exploit the cellular machinery to produce viral proteins. Viral vectors are engineered vectors, in which pathogenic viral genes are replaced by therapeutic gene(s), leaving intact only those sequences that are required for viral vector packaging and transfer. The resulting viruses are able to "transduce" genes of interest into cells with higher efficiency than all others transfection methods [34–36].

Up to now, four derived viral vectors are mostly used to produce particles for infection of neuronal primary cells or NSCs: adenoviruses, adeno-associated viruses, lentiviruses, and herpes simplex viruses-derived vectors. To choose the appropriate vector, the specific characteristics of each vector have to be taken in

account: the cloning capacity, the tropism, and the onset of transgene expression. Despite continuous improvements of viral vectors
and the associated protocols, there are still some critical disadvantages for viral vector use for delivery genes to neuron primary culture. In particular, the production of viral vectors is time consuming
and requires specialized equipment, even if commercial kits are
now available to facilitate the procedure [35, 37–39].

2.1 Adenoviruses

Adenoviruses (AdVs) are non-enveloped icosahedral viruses composed of a nucleocapsid and a double stranded linear DNA genome
[40]. Viral replication and assembly occur only in the nucleus of
infected cells, and virions are released by cellular lysis. There are
three main features that make AdV-derived vectors a tool of choice
for primary neuron transfection: (1) the possibility to transduce
long inserts (up to 36 kilobases), (2) the ability to transduce both
dividing and non-dividing cells and (3) the ability to grow the vectors to very high titers.

In the beginning, their use to transduce primary cell was limited by their immunogenicity. With time though, they have been
engineered to reduce their immunogenic potential by substituting
some viral genes (E1/E2/E3/E4) with genes of interest [33, 41–
43]. However, only the development of helper-dependent adenoviruses (HD-Ads), in which all viral genes except for the two
inverted terminal repeats (ITRs) and the packaging signal (*psi*)
have been deleted, resulted in a minimal immune response and
provided a prolonged transgene expression [44–46]. Thanks to
these features and the large insert capacity of these vectors, it was
possible to generate, for example, a new chronic in vitro model of
Huntington's disease, where the huntingtin cDNA (10 kilobases)
and its mutated form were transduced through AdV vectors in
murine cortical and striatal primary neurons [47].

The major limitation of recombinant adenoviruses is the limited duration of gene expression through a preexisting humoral
and cellular immunity against common human AdVs serotype
[48]. A recent and promising alternative is the use of non-human
adenoviral vectors, like canine adenovirus, as reported by the Alves
group. The helper-dependent canine adenovirus type 2 has an
insert capacity of 30 kilobases, shows a preferential tropism for
human neurons in vitro and leads to a long-term transgene expression with very low cytotoxic effects [49].

**2.2 Adeno-
associated Viruses**

The adeno-associated virus (AAV) is a small icosahedral animal
virus with a single-stranded DNA genome of 4.7 kilobases [50].
Since distinct serotypes of AAV present different capsid proteins,
the tropism of the AAV vector can be specifically selected for the
cells presenting certain cell surface receptors [50]. AAV is not able
to actively replicate itself and generally integrates in a precise region
of human chromosome 19, where it persists in a latent form [51].

A second infection with a helper virus provides all factors necessary for active replication. An engineered AAV version, the derived recombinant adeno-associated viral vectors (rAAV), is completely helper-free and rarely integrates into the genome [51]. This feature, combined with the long-term gene expression in both dividing and non-dividing cells and the low immunogenicity, allowed the engineering of AAV vectors that are safer as well as cell-specific subtypes [37]. The AAV-2 subtype has been the first and most commonly used serotype to deliver genetic material into cells within the CNS. Currently, at least ten distinct AAV serotypes have been identified to infect different neuronal cell types with different efficiencies [34, 37, 52, 53].

To achieve a more effective gene transfer in tissues refractive to rAAV2 transduction, pseudotyping its genome with capsids from alternative serotypes is becoming an increasingly employed practice [54–57] . For instance, to improve the infection of glioma cells in vivo and in vitro, numerous rAAV2 capsid variants have been tested, such as the deletion/modification of three surface-exposed tyrosine residues, which are sites for phosphorylation and subsequent ubiquitination, thus reducing virus proteosomal degradation [58]. Recently, Kotterman and colleagues showed that an AAV2 variant, AAV r3.45, efficiently and selectively transduced adult human, mouse, rat NSCs both in vitro and in vivo [59–61]. These findings suggest AAV r3.45 as a potential tool to investigate neurogenesis in the adult brain and for novel gene therapy approaches and cell replacement therapy applications.

Interestingly, rAAV vectors expressing the human HIF-1α gene were also used to transduce primary cultured hippocampal neurons to demonstrate that this protein significantly reduces apoptosis induced by amyloid-beta protein. These findings suggest gene therapy based on HIF-1α as a potential treatment for neurodegenerative disorders such as Alzheimer's disease [62]. However, a key limitation in rAAV vector application is their low packaging capacity. Different solutions have been proposed for this problem, including in vivo concatemerization of rAAV [63–65]. Recently, Gompf and colleagues proposed another interesting approach that involves the current use of two AAV vectors: the first one encodes for the gene of interest in a double-inverted open reading frame driven by a strong neuron or generalized promoter, and retained in an inverse, nonsense orientation in absence of Cre; the second one delivers a Cre-recombinase under the control of a subtype-specific promoter [66]. Since small amounts of Cre-recombinase are needed to reorient the transgene for the expression, a significantly weaker, but neuronal subtype-specific promoter can drive sufficient Cre expression. This "dual system" overcomes the limitation of the low packaging capacity of rAAV and gives a high and cell type specific expression of the gene of interest.

2.3 Lentiviruses

Lentiviruses are members of the retrovirus family. They are enveloped particles containing homodimers of a linear single-stranded RNA genome [67]. Lentiviruses are the only members of the retrovirus family able to enter the host nuclei of non-dividing cells and stably integrate into their genome, allowing the persistent expression of the gene of interest. Lentiviral vectors (LV) have a relatively large cloning capacity of around 9 kilobases and are derived from HIV, the most extensively studied lentivirus [68]. To increase the biosafety and efficiency of transfection, three generations of LV have been developed. The third generation of LV is replication defective and self-inactivating. In addition, the packaging elements have been split into four different plasmids [38, 69, 70]. Moreover, to avoid any possibility of integration, when not needed, integration-deficient LV have been generated carrying an inactive integrase protein [71]. A significant improvement to increase the transduction potency and the range of targetable tissues of HIV-based vectors is derived from the incorporation of the envelope proteins of other viruses in HIV particles. For example, the G glycoprotein of the vesicular stomatitis virus was the first widely used heterologous envelope protein pseudotyping HIV [69, 72]. More recently, sophisticated modifications of the envelope were reported allowing the specific transduction of neurons in vitro and in vivo [72–76]. For instance, Eleftheriadou and colleagues generated a novel LV by co-expressing into the lentiviral surface a fusogenic glycoprotein composed of mutated sindbis G and an antibody against a cell-surface receptor on the presynaptic terminal of the neuromuscular junction (Thy1.1, p75[NTR], or coxsackie virus and adenovirus receptor). By using these vectors, the authors obtained a specific and efficient transduction of motor neurons in vitro and in vivo. [77]. LVs have also been successfully used to transduce many neuronal cell types in vivo, including adult neuronal stem cells, oligodendrocytes, and astrocytes [38]. For instance, LVs were used to infect cells of the dorsal root ganglion (DRG) with protein kinase A (PKAc) and green fluorescent protein (GFP) to test whether the activation of RhoA after spinal cord injury could be inhibited by PKAc expression in the DRG. The results showed that the LV could mediate exogenous PKAc gene expression in the DRG and could thus inhibit spinal cord injury-induced RhoA activation by delivery of LV/PKAc-IRES-GFP through the sciatic nerve [78]. A lentiviral-based gene therapy was also recently employed in a clinical trial for Parkinson's disease [79]. In this trial, ProSavin, a lentiviral vector engineered for gene therapy, was used to restore local and continuous dopamine production in 15 patients with advanced Parkinson's disease. ProSavin was observed to be safe and well tolerated in these patients. In addition, a marked improvement in motor behavior was observed in all treated patients.

2.4 Herpes Simplex Viruses

Herpes simplex virus (HSV-1), which is characterized by a natural tropism for neurons, was the first virus utilized for high efficiency gene delivery into neurons. Beyond the specificity for neuronal cells, another feature that places HSVs high on the list of choices for transfecting primary neuronal cultures is its ability to be transported and transferred across synapses in a retrograde matter, that can be used to trace connections within neuronal networks [80]. To facilitate the use of HSV-1 vectors in experiments involving neuronal cells, recombinant HSV-1 based vectors have been developed [81]. Amplicons are replication-incompetent helper-dependent vectors derived from HSV-1, with several advantages: they do not encode any viral protein and are therefore neither toxic for the infected cells nor pathogenic for the inoculated animals, and elicit low levels of adaptive immune responses. In addition, due to the large genome capacity of HSV-1 (up to 150 kilobases), it is possible to transduce entire genes including regulatory sequences, which allows the observation of more specific gene expression effects and dynamics. Moreover, since the viral genome does not integrate into cellular chromosomes, there is a low probability to induce insertional mutations [39].

An interesting application involving HSV-based vectors is the transduction of NSCs with the aim to deliver thymidine kinase (tk) into murine glioma cells combined with the administration of the apoptotic prodrug Glancicovir [82]. HSV/tk-transduced NSCs efficiently infiltrated into the tumor providing a promising option to treat patients. However, a major drawback of the use of HSV-based vectors is the difficult production of high-titer stocks of particles free of helper viruses that may lead to cytotoxic effects and/or immune responses [81]. Since the expression of these amplicons peaks at 3–4 days after transduction, they are often used for experiments needing a short and transient expression of the inserted construct. Friedman and colleagues took advantage of this feature and recently reported an optimized protocol to study functions of neuronal genes cloned into HSV vectors after ex vivo transduction of rodent brain slices [83].

3 Nonviral Gene Transfer

3.1 Physical Methods

3.1.1 Electrical Transfection

The term "electroporation" describes the transfection techniques where plasmid constructs or oligonucleotides are delivered inside cells through the application of an electrical field. This induces an abrupt increase of transmembrane potential, resulting in structural rearrangements of the cell membrane and in the formation of transient, metastable aqueous pores of varying diameters. These pores allow exogenous material, such as charged DNA, to electro-diffuse along the electrical field from the medium into the cytoplasm [84].

A major advantage of electroporation is the possibility to simultaneously introduce different expression constructs and of varying sizes inside cells, thus broadening the spectrum of the possibilities available to investigate cells and allowing for instance the easy co-transfection of reporter proteins, such as GFP [33]. Another interesting aspect of electroporation is that it can be used not only to transfer plasmids, but also to potentially target any charged macromolecule inside neurons, including dyes, drugs, antibodies, antisense oligonucleotides, double-stranded RNAs, and bacterial or yeast artificial chromosomes [85]. Different cell types showed a specific capacity to undergo and recover from electroporation, and among them, neurons have proven to be a challenging subject for this procedure. Indeed, the transfection efficiency with electroporation ranges between 10 and 20%. [86]. Therefore, different variants of this technique have been successfully developed such as single-cell-electroporation (SCE) and nucleofection.

SCE allows the transfer of expression plasmids or siRNA molecules into individual cells in vitro and in vivo [87–89]. In this method, a target neuron in vivo is identified and visualized in the intact brain by two-photon microscopy and subsequently electroporated with a high-resistance patch pipette containing the plasmid DNA. Transfected neurons can afterwards be subjected to whole cell patch-clamp recordings and imaging experiments [89]. It is therefore possible to analyze the functional integration of individual neurons within a native neuronal network, to perform silencing-rescue experiments and to monitor the development as well as physiological or pathological functions of single neurons [88]. For instance, Kabakov and colleagues used SCE to transfect alpha-CAMKII, a wild-type and a mutated form in its dead domain K42M, into organotypic hippocampal slice cultures. They were able to observe the behavior of two neighboring neurons expressing either the wild-type or the mutant CAMKII transfected under the same conditions [90]. In addition, Pagès and colleagues recently indicated SCE as a potential alternative to transfect adult cortical neurons and subsequently image fluorescently tagged synaptic proteins over several days to weeks in vivo [91].

A further development to classical electroporation techniques is nucleofection, a "nucleus-specific" variant of electroporation. Several features of nucleofection make it the nonviral method of choice to achieve comparatively high levels of transfection in cultured primary neurons. In contrast to conventional electroporation, nucleofection combines specific electrical parameters with reagents optimized for specific cell types. This is aimed at ensuring a high viability of the nucleofected cells. In nucleofection, strong electric fields are used to deliver DNA or other biologically active molecules into the nucleus in a cell division-independent manner. The cell type-specific, high-voltage pulses generated by the devices used in nucleofection facilitate the penetration of foreign molecules into the nucleus, possibly by creating transient holes in the nuclear envelope or otherwise rendering the nuclear pores more permeable to large molecules. Nucleofection thus allows the transfection of a wide variety of primary cells and cell lines, including ones that divide slowly or are mitotically inactive. The short, high-voltage pulses and specific buffers used during the procedure also diminish the generation of cytotoxic anions typical of conventional electroporation with longer-lasting pulses. These modifications result in lower cell mortality and higher transfection efficiencies. As such, nucleofection has achieved transfection efficiencies of over 50–60% for numerous primary neuronal cell cultures, such as cortical, hippocampal, striatal, cerebellar, retinal ganglion, DRG neurons and NSCs (Fig. 1) [92–94]. For instance, cortical neurons were successfully nucleofected with Huntingtin associated protein 1 to demonstrate its role in regulating the sorting gamma-aminobutyric acid type A receptor and consequently in the development and maintenance of inhibitory synapses [95]. Qing-Jian Han and colleagues also used nucleofection to upregulate or downregulate the expression of endogenous inhibitor 5 of protein phosphatase 1 in dissociated DRG neurons [96].

Recently developed nucleofector devices also offer the possibility to achieve higher transfection efficiencies using a lower number of cells and without increasing cell mortality. For example, Bertram and colleagues reported a great improvement of their transfections of mouse cortical NSCs by using these devices and achieved a transfection rate over 80%, with a mortality of less than 30% [97]. Moreover, in comparison with first generation instruments, the use of new nucleofector devices does not affect the differentiation behavior of mouse cortical NSC. McCall and colleagues optimized the protocol for the nucleofection of adult rat DRG with large plasmids—useful, for example, for large viral vectors or genes with their regulatory elements—without adding Fetal Bovine Serum (FBS) to the medium after electroporation, since this can affect the studies on regulators of survival, neurite growth, and transcriptional response [98]. According to their protocol, a transfection efficiency of 39–42% was achieved.

Another improvement of the nucleofection device of interest for certain applications is the 96-well shuttle system, which allows the simultaneous testing of up to 96 different plasmids, siRNA, miRNA, and/or conditions [99]. However, so far, the transfection efficiencies achieved with the 96-well shuttle system are generally lower than those obtained with the nucleofection variants discussed above. Hutson and colleagues described several improvements of the protocol for the transfection of postnatal rat cerebellar granule neurons (CGNs) with these devices [100]. Under optimal conditions they achieved a 28% transfection efficiency and 51% viability for these cells.

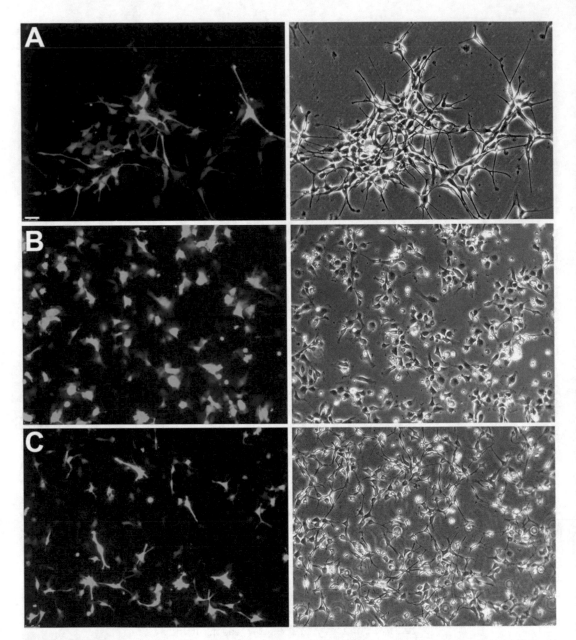

Fig. 1 Neuronal Stem Cells nucleofected with expression plasmid encoding green fluorescent protein (GFP). (**a**) Neural stem cells derived from human fetal brain (human fetal NSCs); (**b**) neural progenitors derived from human induced pluripotent stem cells (human IPSC-derived NPCs); and (**c**) neural stem cells derived from mouse embryonic stem cells (mouse ESC-derived NSCs). Cells were nucleofected with a GFP reporter vector using Nucleofector 2b™ (Amaxa). The images were acquired 48 h after nucleofection and analyzed by fluorescence (*left panels*) and phase contrast microscopy (*right panels*). Scale bar = 100 μm. (Courteously provided by Jacopo Zasso and Luciano Conti)

The main drawback of a nucleofection of primary neurons is that this method can be applied only to cells in suspension [92–94]. Thus, primary neurons have to be transfected immediately after their dissociation, hence impeding experiments on mature neurons. This relevant limitation may be somehow overcome by using "new era" devices (Y unit 4D-Nucleofector system, 24-well culture plates, Lonza) that allow the nucleofection of adherent neuronal cells at any developmental stage with high efficiency and viability [101, 102].

3.1.2 Magnetofection

Magnetofection is a simple and highly efficient method to transfect cells in vitro and in vivo. This technology utilizes magnetic force to trigger a rapid accumulation of nucleic acid-associated magnetic particles around the cell, thus promoting their uptake. This technique was initially developed mainly to improve gene transfer in cell cultures, but it was progressively adapted for use with difficult-to-transfect cells, such as primary neurons [103, 104]. Specific magnetic nanoparticles called "NeuroMag" were formulated to achieve high transfection efficiencies for these cells [105]. NeuroMag has been developed to efficiently transfect different types of nucleic acids, such as DNA, RNA, or oligonucleotides in serum compatible condition, for both transient and stable transfection.

There are some important characteristics that make magnetofection particularly useful for the transfection or transduction of primary neurons. Unlike electroporation methods, cells can be in an adherent state during transfection, hence also allowing the transfection of neurons at the later stages of differentiation or multiple transfections of the same neuronal culture at different time points. Moreover, NeuroMag nanoparticles form electrostatic and hydrophobic interactions with nucleic acids, thus being fully biodegradable and do not interfere with cellular homeostasis. NeuroMag has been verified to be extremely useful for transfecting different primary neuronal populations, such as dopaminergic, hippocampal, and cortical neurons [106–112]. Fallini and colleagues exploited the possibility of applying magnetofection to transfect isolated primary motor neurons, a suitable in vitro model to study motor neuron degeneration and diseases [113–115]. A transfection efficiency of higher than 45% was obtained without significant alterations of the cells' morphology or the survival of the motor neurons. Importantly, NeuroMag nanoparticles allowed obtaining an efficient co-transfection of reporter genes together with a plasmid encoding the Survival of Motor Neuron (SMN) protein for overexpression experiments or an antisense construct to knock down its expression. For the first time, the active transport of SMN-containing granules anterogradely and retrogradely along motor neurons axons was demonstrated, supporting a role for SMN in the regulation of mRNA localization and axonal transport through its interaction with RBPs such as IMP1 [113, 114].

NeuroMag-mediated transfection has also been proposed as a suitable method to transfect NSCs and therefore to carry out high-throughput projects for the discovery of new regulators of neuronal differentiation [116]. Pickard and colleagues have recently reported an optimization of the protocol to magnetofect NSCs [117]. This group showed that the combination of an oscillating magnetic field with a monolayered neuronal culture yielded the highest transfection efficacy (32.2%) for NSC magnetofection without significantly interfering with cell viability. This protocol thus offers a valid nonviral alternative for the genetic modification of this important neural cell transplant population.

3.1.3 Microinjection

The microinjection method consists of the injection of nucleic acids directly into a cell's cytoplasm or nucleus with thin and high-resistance glass capillaries. This technique is useful when a high level of protein expression is required, since it allows the introduction of plasmids directly in the nucleus. Notably, it allows the simultaneous and reproducible injection of different cDNAs in defined proportions [118]. For instance, Bounhar and colleagues microinjected Bax cDNA with either prion protein or Bcl2 cDNA into cultured human primary neurons and demonstrated that the cytosolic prion protein protects human neurons against Bax-mediated apoptosis [119, 120]. In addition, microinjection permits injecting molecules that cells cannot produce and targeting them to specific compartments [121, 122]. For instance, Kole and colleagues microinjected holocytochrome c (heme-attached) protein directly into the cytoplasm of mouse embryonic fibroblasts or primary sympathetic neurons [123]. The reported protocol is useful to investigate apoptosis but it could also be adapted for the microinjection of other proteins of interest. Moreover, microinjection is well suited for single cell studies in experiments requiring specific gene transfer to a single, identified neuron [124], precise control of neuronal co-transfection of two or more plasmid DNAs [118, 125], or electrophysiological recordings and optical imaging.

On the other hand, microinjection is not ideal for biochemical assays requiring a large number of cells and higher transfection efficiencies (i.e., greater than 10–20%). Another complication of this technique is the stress induced by the penetration of the plasma membrane by injection capillary, which, depending on the cells analyzed, may lead to a very low survival rates. It is therefore critical to ensure that cell viability is not compromised in microinjection experiments by including appropriate controls [126]. Another concern is that microinjection requires a high degree of manual expertise by the operator to inject large numbers of neurons in a short period of time. The use of fully automated systems is progressively overcoming this issue, resulting in a correct cell detection rate of up to 87% and a successful marker delivery rate of up to 67.5% [124]. These results demonstrate that the new automated systems are capable of better performance than expert operators, suggesting a potential for large-scale applications.

3.1.4 Biolistics

Biolistic transfection is a physical method that utilizes the rapid propulsion of small metal (e.g., gold) particles to introduce their coating of plasmid DNA into target cells or tissues. Given its physical basis, this method depends neither on the biochemical features of cell membranes nor on the growth rate of the cells [127, 128]. It was initially developed as a method of gene transfer into monocotyledonous plants, since the presence of cell walls made transfection with preexisting methods difficult [129, 130]. However, it is now becoming a valuable tool to transfect cultured neural cells, tissue slices, or living organs as well, facilitating experiments on individual neurons in the context of an entire brain or spinal cord [131, 132].

In particular, biolistic transfection has become the method of choice for transfecting neurons in slice cultures because individual transfected neurons can be examined in isolation, often in conjunction with two-photon microscopy. Two successful examples of the application of this technique with organotypic hippocampal slices are reported by Carta and colleagues, who identified CaMKII-dependent phosphorylation of GluK5 as responsible for kainate receptor plasticity [133] and by Oliveira and colleagues, who demonstrated that neurofibromin is the major Ras inactivator in dendritic spines [134]. In another study on epilepsy syndrome, Lachance-Touchette and colleagues suggested biolistic transfection as an appropriate tool to perform single cell manipulation in coronal brain slices of the occipital cortex for studying the effects of different mutations of $GABA_A$ alpha1 subunit on both dendritic spine and GABAergic bouton formation [135].

The major limitation of using biolistics is the cells damage due to high gas pressure and to the size of metal particles [136]. O'Brien and colleagues reported several improvements to overcome these important drawbacks [131, 136–138].Recently, they demonstrated that the use of custom made low-pressure barrel and smaller DNA-coated nanoparticles (40 nm) minimize tissue damage compared to larger traditional gold particles (1 μm), in organotypic brain slices [138]. In addition, the slices transfected with 40 nm particles displayed a significantly larger number of viable and transfected cells (about 32%, 5 day post-transfection) and expressed the exogenous proteins for many weeks.

3.2 Chemical Transfection Methods

3.2.1 Lipofection

Lipofection-based transfection utilizes synthetic cationic lipid molecules (for example Lipofectamine 2000, TransIT) to deliver nucleic acids into cells [139]. Liposomes are often combined with neutral helper lipids that participate to form small unilamellar vesicles that interact spontaneously with negatively charged nucleic acids and ease the fusion of these complexes with the cells' plasma membrane [139]. Cationic liposomes have been demonstrated to have high but variable transfection efficiencies in dividing cells and a good efficiency in post-mitotic neurons (30–50%) [140, 141]. For instance, this technique was successfully used to overexpress an mCherry-tagged TWIK-related spinal cord K$^+$ channels (TRESK) subunit in mouse trigeminal ganglion (TG) neurons [142]. Since a TRESK-specific channel opener is not currently available, the resulting twofold increase of the protein inside the transfected cells was sufficient to mimic its effects in small-diameter TG neurons and to indicate channel openers as potential therapeutics for the treatment of migraine and other chronic pain symptoms [142]. A transfection efficiency of 25% was also measured for DRG neurons, by combining the use of cationic liposomes with mRNA molecules instead of cDNA. Since mRNA translation takes place in the cytosol, protein expression from exogenous mRNA does not require entry of exogenous plasmids into the cell's nucleus. For this reason, this approach has the potential to be an efficient method for heterologous expression in post-mitotic cells, such as primary neurons [143]. The same approach has been successfully used to transfect NSCs derived from the subventricular zone of adult C57BL/6 mouse brain with Lipofectamine 2000 or TransIT [144]. Interestingly, the use of cationic liposomes yielded transfection efficiencies of up 40–50% while the electroporation efficiency was 60–70% with a lower expression of protein per cell. An optimization of the protocol (60% transfection efficiency) was obtained with lower mRNA lipoplex concentration in small neurospheres plated onto extracellular matrix and cultured as adherent monolayers [144].

A second generation of lipofection reagents, however, utilizes non-liposomal lipids (Fugene or Effectene) to form complexes with nucleic acids. An efficiency of 5.1% for primary glia [145] and 2.4% for mesencephalic neurons [146] was shown. Furthermore, Fugene is reported to be a suitable transfection reagent for adult hippocampal progenitors (AHP). After screening six nonviral agents, Fugene showed a lower cytotoxicity without perturbing the viability or rate of proliferation of treated cells. A long-term expression of lipofected plasmids can also be achieved by selecting for random transgene integration [147].

In addition, lipofection reagents efficiently transfer siRNAs, microRNAs, and other oligonucleotides into post-mitotic neurons, with efficiencies of up to 83% in primary rat hippocampal, cerebellar granule and cortical neurons [148–150]. For instance, Lipofectamine 2000 was successfully used to transfect primary cortical neurons with siRNA for Sirt3 protein for 72 h obtaining a decrease greater than 80% of Sirt3 expression without significantly affecting cell viability [150].

3.2.2 Calcium
Phosphate/DNA
Co-precipitation

Transfection by calcium phosphate/DNA co-precipitation is the first transfection method historically used to treat primary neuronal cultures and is one of the most established and characterized methods to transfect different neuronal populations as well as cell lines in vitro [151–153]. It is widely used for its ease and low costs, but calcium phospate/DNA co-precipitation has become restricted to specific applications because of its comparatively low transfection efficiency. This procedure remained for a long time a commonly employed protocol mainly because of the possibility to transfect neurons at all stages of differentiation, including cells that had already formed a complex network. A key advantage of this method is that it is easy to regulate the expression of the inserted constructs by titrating the concentrations of DNA put in solution with calcium phosphate [154].

The technique is based on the interaction between DNA and calcium phosphate, which, when in solution, form a precipitate that maintains the properties of the nucleic acid. Upon addition to a culture medium, this precipitate settles onto the cells and is be up-taken by endocytosis, thus transferring the genetic material into the cells. In post-mitotic neurons entry into the nucleus is more difficult, so expression rates and transfection efficiencies are reduced (typically between 1 and 5%). To overcome this bottleneck several approaches to optimize transfection by calcium phosphate/DNA co-precipitation have been proposed, especially with an emphasis on avoiding cytotoxicity, which tends to increase with as transfection rates rise with this method. In particular, the determination of (1) the optimal pH for precipitate formation and culture medium, (2) the best procedure to mix calcium phosphate/DNA solution to generate smaller precipitates, and (3) the optimal incubation time of the cells with co-precipitates has resulted in significant improvements of transfection efficiencies (up 50%) [152, 155, 156]. These recent developments indicate that transfection protocols using calcium phosphate/DNA co-precipitation might experience a comeback as one of the most cost efficient and effective transfection techniques for cultured neuronal cells, especially primary neurons and therefore, it could be increased the spectrum of experiments where to exploit this technique.

4 Conclusion

The wide range of methods for neuronal gene transfer discussed in this chapter allows routine transfection of most neuronal preparations including dissociated cultured neurons, acute and organotypic brain slices, neurons in vivo, and NSCs. At present, only viral methods achieve very high transfection efficiencies (over 95%) of post-mitotic neurons. The engineered viral vectors discussed here permit the delivery of a broad range of inserts with size of up to 150 kilobases. It is thus possible to transfect very large proteins or proteins with their regulatory elements [34–39]. In addition, for the vectors with low insert capacity, such as AAV, intriguing solution have been proposed to overcome this limitation [63–66]. Moreover, it is also possible to choose the viral vectors that allow a desired onset of expression raging from a few days (HSV) to stable, long-term expression (LV). Other important solutions have been also proposed to obtain a more specific tropism. The "dual system" proposed for the AAV virus and the use of fusogenic proteins reported for LV seem very promising solutions [66, 77].

Despite these benefits, approaches using viral vectors still have significant limitations. The use of viral vectors requires biosafety 2 level equipment and the production of viral particles is still laborious and time consuming. In addition, the immune response that may be stimulated by these vectors remains a concern, especially in the context of gene therapy. For these reasons, much effort has been aimed at developing nonviral methods, such as nucleofection, magnetofection, and lipofection, which also attain significant transfection efficiencies. These three approaches now allow to carry out quantitative biochemical analyses either in knocked down or overexpression experiments [95, 97, 113, 142, 143]. Moreover, they can be performed in 96-well plate formats and making it possible to simultaneously test different conditions, drugs, plasmids, or siRNAs. These techniques are fast, easy to perform, cost-effective and require only basic equipment. Moreover, they show low cytotoxicity and tend not to interfere significantly either with proliferation or the differentiation of NSCs [97, 116, 117, 147].

Despite important advances in the past years, significant progress is constantly reported. Concerning nucleofection, the production of new devices that give the opportunity to transfect cells in adherent states could overcome one of the major limitations of this technique. With magnetofection, the testing of different magnetic fields (static or oscillatory) seems to improve this procedure [117]. Also for lipofection, new chemical reagents and protocols are constantly developed to ensure a minimum cytotoxicity and better transfection efficiencies [157].

By microinjection and biolistics it not possible to easily achieve high transfection efficiencies; instead, they are more suitable to introducing molecules that cells cannot produce and to target them to specific subcellular compartments, or to transfect individual neurons in intact brain slices [123, 133, 135]. The development of new, fully automated devices for microinjection, however, could make large-scale transfection approaches possible also for this technique [158].

Acknowledgments

We thank Dr. Anna Cereseto, Dr. Luciano Conti, and Martin Michael Hanczyc for their comments on the manuscript. We apologize to all authors whose work could not be cited due to space restrictions. This work was supported by the MaDEleNA Project funded by PAT (Autonomous Province of Trento, Italy).

References

1. Millet LJ, Gillette MU (2012) Over a century of neuron culture: from the hanging drop to microfluidic devices. Yale J Biol Med 85:501–521

2. Kobayashi M, Kim J-Y, Camarena V, Roehm PC, Chao MV, Wilson AC, Mohr I (2012) A primary neuron culture system for the study of herpes simplex virus latency and reactivation. J Vis Exp pii:3823. doi:10.3791/3823

3. Silver I, Deas J, Erecińska M (1997) Ion homeostasis in brain cells: differences in intracellular ion responses to energy limitation between cultured neurons and glial cells. Neuroscience 78:589–601. doi:10.1016/S0306-4522(96)00600-8

4. Bird CW, Gardiner AS, Bolognani F, Tanner DC, Chen C-Y, Lin W-J, Yoo S, Twiss JL, Perrone-Bizzozero N (2013) KSRP modulation of GAP-43 mRNA stability restricts axonal outgrowth in embryonic hippocampal neurons. PLoS One 8:e79255. doi:10.1371/journal.pone.0079255

5. Moroz LL, Kohn AB (2013) Single-neuron transcriptome and methylome sequencing for epigenomic analysis of aging. Methods Mol Biol 1048:323–352. doi:10.1007/978-1-62703-556-9_21

6. Gaven F, Marin P, Claeysen S (2014) Primary culture of mouse dopaminergic neurons. J Vis Exp 91:e51751. doi:10.3791/51751

7. Ray B, Chopra N, Long JM, Lahiri DK (2014) Human primary mixed brain cultures: preparation, differentiation, characterization and application to neuroscience research. Mol Brain 7:63. doi:10.1186/s13041-014-0063-0

8. Brelstaff J, Ossola B, Neher JJ, Klingstedt T, Nilsson KPR, Goedert M, Spillantini MG, Tolkovsky AM (2015) The fluorescent pentameric oligothiophene pFTAA identifies filamentous tau in live neurons cultured from adult P301S tau mice. Front Neurosci 9:184. doi:10.3389/fnins.2015.00184

9. Harrill JA, Chen H, Streifel KM, Yang D, Mundy WR, Lein PJ (2015) Ontogeny of biochemical, morphological and functional parameters of synaptogenesis in primary cultures of rat hippocampal and cortical neurons. Mol Brain 8:10. doi:10.1186/s13041-015-0099-9

10. Sendrowski K, Sobaniec W, Stasiak-Barmuta A, Sobaniec P, Popko J (2015) Study of the protective effects of nootropic agents against neuronal damage induced by amyloid-beta (fragment 25-35) in cultured hippocampal neurons. Pharmacol Rep 67:326–331. doi:10.1016/j.pharep.2014.09.013

11. Giordano G, Costa LG (2011) Primary neurons in culture and neuronal cell lines for in vitro neurotoxicological studies. Methods Mol Biol 758:13–27. doi:10.1007/978-1-61779-170-3_2

12. Capela JP, da Costa AS, Costa VM, Ruscher K, Fernandes E, Bastos Mde L, Dirnagl U, Meisel A, Carvalho F (2013) The neurotoxicity of hallucinogenic amphetamines in primary cultures of hippocampal neurons. Neurotoxicology 34:254–263. doi:10.1016/j.neuro.2012.09.005

13. Vilela LR, Gobira PH, Viana TG, Medeiros DC, Ferreira-Vieira TH, Doria JG, Rodrigues F, Aguiar DC, Pereira GS, Massessini AR, Ribeiro FM, de Oliveira ACP, Moraes MFD, Moreira FA (2015) Enhancement of endocannabinoid signaling protects against cocaine-induced neurotoxicity. Toxicol Appl Pharmacol 286:178–187. doi:10.1016/j.taap.2015.04.013

14. Yang N, Ng YH, Pang ZP, Südhof TC, Wernig M (2011) Induced neuronal cells: how to make and define a neuron. Cell Stem Cell 9:517–525. doi:10.1016/j.stem.2011.11.015

15. Bailey JA, Lahiri DK (2006) Neuronal differentiation is accompanied by increased levels of SNAP-25 protein in fetal rat primary cortical neurons: implications in neuronal plasticity and Alzheimer's disease. Ann N Y Acad Sci 1086:54–65. doi:10.1196/annals.1377.001

16. Yang J, Ruchti E, Petit J-M, Jourdain P, Grenningloh G, Allaman I, Magistretti PJ (2014) Lactate promotes plasticity gene

expression by potentiating NMDA signaling in neurons. Proc Natl Acad Sci U S A 111:12228–12233. doi:10.1073/pnas.1322912111

17. May PC, Boggs LN, Fuson KS (1993) Neurotoxicity of human amylin in rat primary hippocampal cultures: similarity to Alzheimer's disease amyloid-? Neurotoxicity. J Neurochem 61:2330–2333. doi:10.1111/j.1471-4159.1993.tb07480.x

18. Murphy DD, Rueter SM, Trojanowski JQ, Lee VM-Y (2000) Synucleins are developmentally expressed, and alpha -Synuclein regulates the size of the presynaptic vesicular pool in primary hippocampal neurons. J Neurosci 20:3214–3220

19. Krantic S, Isorce N, Mechawar N, Davoli MA, Vignault E, Albuquerque M, Chabot J-G, Moyse E, Chauvin J-P, Aubert I, McLaurin J, Quirion R (2012) Hippocampal GABAergic neurons are susceptible to amyloid-β toxicity in vitro and are decreased in number in the Alzheimer's disease TgCRND8 mouse model. J Alzheimers Dis 29:293–308. doi:10.3233/JAD-2011-110830

20. Popugaeva E, Pchitskaya E, Speshilova A, Alexandrov S, Zhang H, Vlasova O, Bezprozvanny I (2015) STIM2 protects hippocampal mushroom spines from amyloid synaptotoxicity. Mol Neurodegener 10:37. doi:10.1186/s13024-015-0034-7

21. Li XJ, Valadez AV, Zuo P, Nie Z (2012) Microfluidic 3D cell culture: potential application for tissue-based bioassays. Bioanalysis 4:1509–1525. doi:10.4155/bio.12.133

22. Jacobs BM (2015) A dangerous method? The use of induced pluripotent stem cells as a model for schizophrenia. Schizophr Res 168:563–568. doi:10.1016/j.schres.2015.07.005

23. Kanzaki S (2014) Gene and drug delivery system and potential treatment into inner ear for protection and regeneration. Front Pharmacol 5:222. doi:10.3389/fphar.2014.00222

24. Parr-Brownlie LC, Bosch-Bouju C, Schoderboeck L, Sizemore RJ, Abraham WC, Hughes SM (2015) Lentiviral vectors as tools to understand central nervous system biology in mammalian model organisms. Front Mol Neurosci 8:1–12. doi:10.3389/fnmol.2015.00014

25. De la Rossa A, Jabaudon D (2015) In vivo rapid gene delivery into postmitotic neocortical neurons using iontoporation. Nat Protoc 10:25–32. doi:10.1038/nprot.2015.001

26. Chen X, Zhao X, Zhang M, Wei S (2015) Nuclear respiratory factor-2α and adenosine triphosphate synapses in rat primary cortical neuron cultures: the key role of adenosine monophosphate-activated protein kinase. Mol Med Rep. doi:10.3892/mmr.2015.4140

27. Del Pino J, Frejo MT, Baselga MJA, Capo MA, Moyano P, García JM, Díaz MJ (2015) Neuroprotective or neurotoxic effects of 4-aminopyridine mediated by KChIP1 regulation through adjustment of Kv 4.3 potassium channels expression and GABA-mediated transmission in primary hippocampal cells. Toxicology 333:107–117. doi:10.1016/j.tox.2015.04.013

28. Sato T, Ishikawa M, Mochizuki M, Ohta M, Ohkura M, Nakai J, Takamatsu N, Yoshioka K (2015) JSAP1/JIP3 and JLP regulate kinesin-1-dependent axonal transport to prevent neuronal degeneration. Cell Death Differ 22:1260–1274. doi:10.1038/cdd.2014.207

29. Oh S-M, Chang M-Y, Song J-J, Rhee Y-H, Joe E-H, Lee H-S, Yi S-H, Lee S-H (2015) Combined Nurr1 and Foxa2 roles in the therapy of Parkinson's disease. EMBO Mol Med 7:510–525. doi:10.15252/emmm.201404610

30. Klein SM, Behrstock S, McHugh J, Hoffmann K, Wallace K, Suzuki M, Aebischer P, Svendsen CN (2005) GDNF delivery using human neural progenitor cells in a rat model of ALS. Hum Gene Ther 16:509–521. doi:10.1089/hum.2005.16.509

31. Mariotti V, Greco SJ, Mohan RD, Nahas GR, Rameshwar P (2014) Stem cell in alternative treatments for brain tumors: potential for gene delivery. Mol Cell Ther 2:24. doi:10.1186/2052-8426-2-24

32. Kim YC, Shim JW, Oh YJ, Son H, Lee YS, Lee SH (2002) Co-transfection with cDNA encoding the Bcl family of anti-apoptotic proteins improves the efficiency of transfection in primary fetal neural stem cells. J Neurosci Methods 117:153–158

33. Washbourne P, McAllister AK (2002) Techniques for gene transfer into neurons. Curr Opin Neurobiol 12:566–573. doi:10.1016/S0959-4388(02)00365-3

34. Grimm D, Kay MA (2003) From virus evolution to vector revolution: use of naturally occurring serotypes of adeno-associated virus (AAV) as novel vectors for human gene therapy. Curr Gene Ther 3:281–304

35. Lentz TB, Gray SJ, Samulski RJ (2012) Viral vectors for gene delivery to the central nervous system. Neurobiol Dis 48:179–188. doi:10.1016/j.nbd.2011.09.014

36. Kim TK, Eberwine JH (2010) Mammalian cell transfection: the present and the future. Anal Bioanal Chem 397:3173–3178. doi:10.1007/s00216-010-3821-6

37. Ojala DS, Amara DP, Schaffer DV (2015) Adeno-associated virus vectors and neurological gene therapy. Neuroscientist 21:84–98. doi:10.1177/1073858414521870

38. Parr-Brownlie LC, Bosch-Bouju C, Schoderboeck L, Sizemore RJ, Abraham WC, Hughes SM (2015) Lentiviral vectors as tools to understand central nervous system biology in mammalian model organisms. Front Mol Neurosci 8:14. doi:10.3389/fnmol.2015.00014

39. Jerusalinsky D, Baez MV, Epstein AL (2012) Herpes simplex virus type 1-based amplicon vectors for fundamental research in neurosciences and gene therapy of neurological diseases. J Physiol Paris 106:2–11. doi:10.1016/j.jphysparis.2011.11.003

40. Wold WSM, Toth K (2013) Adenovirus vectors for gene therapy, vaccination and cancer gene therapy. Curr Gene Ther 13:421–433

41. Lusky M, Christ M, Rittner K, Dieterle A, Dreyer D, Mourot B, Schultz H, Stoeckel F, Pavirani A, Mehtali M (1998) In vitro and in vivo biology of recombinant adenovirus vectors with E1, E1/E2A, or E1/E4 deleted. J Virol 72:2022–2032

42. O'Neal WK, Zhou H, Morral N, Aguilar-Cordova E, Pestaner J, Langston C, Mull B, Wang Y, Beaudet AL, Lee B (1998) Toxicological comparison of E2a-deleted and first-generation adenoviral vectors expressing alpha1-antitrypsin after systemic delivery. Hum Gene Ther 9:1587–1598. doi:10.1089/hum.1998.9.11-1587

43. Andrews JL, Kadan MJ, Gorziglia MI, Kaleko M, Connelly S (2001) Generation and characterization of E1/E2a/E3/E4-deficient adenoviral vectors encoding human factor VIII. Mol Ther 3:329–336. doi:10.1006/mthe.2001.0264

44. Morsy MA, Caskey CT (1999) Expanded-capacity adenoviral vectors--the helper-dependent vectors. Mol Med Today 5:18–24

45. Lowenstein PR, Mandel RJ, Xiong W-D, Kroeger K, Castro MG (2007) Immune responses to adenovirus and adeno-associated vectors used for gene therapy of brain diseases: the role of immunological synapses in understanding the cell biology of neuroimmune interactions. Curr Gene Ther 7:347–360

46. Capasso C, Garofalo M, Hirvinen M, Cerullo V (2014) The evolution of adenoviral vectors through genetic and chemical surface modifications. Viruses 6:832–855. doi:10.3390/v6020832

47. Dong X, Zong S, Witting A, Lindenberg KS, Kochanek S, Huang B (2012) Adenovirus vector-based in vitro neuronal cell model for Huntington's disease with human disease-like differential aggregation and degeneration. J Gene Med 14:468–481. doi:10.1002/jgm.2641

48. Lopez-Gordo E, Podgorski II, Downes N, Alemany R (2014) Circumventing antivector immunity: potential use of nonhuman adenoviral vectors. Hum Gene Ther 25:285–300. doi:10.1089/hum.2013.228

49. Simão D, Pinto C, Fernandes P, Peddie CJ, Piersanti S, Collinson LM, Salinas S, Saggio I, Schiavo G, Kremer EJ, Brito C, Alves PM (2015) Evaluation of helper-dependent canine adenovirus vectors in a 3D human CNS model. Gene Ther. doi:10.1038/gt.2015.75

50. Büning H, Perabo L, Coutelle O, Quadt-Humme S, Hallek M (2008) Recent developments in adeno-associated virus vector technology. J Gene Med 10:717–733. doi:10.1002/jgm.1205

51. Vasileva A, Jessberger R (2005) Precise hit: adeno-associated virus in gene targeting. Nat Rev Microbiol 3:837–847. doi:10.1038/nrmicro1266

52. Rabinowitz JE, Rolling F, Li C, Conrath H, Xiao W, Xiao X, Samulski RJ (2002) Cross-packaging of a single adeno-associated virus (AAV) type 2 vector genome into multiple AAV serotypes enables transduction with broad specificity. J Virol 76:791–801

53. Gao G-P, Alvira MR, Wang L, Calcedo R, Johnston J, Wilson JM (2002) Novel adeno-associated viruses from rhesus monkeys as vectors for human gene therapy. Proc Natl Acad Sci U S A 99:11854–11859. doi:10.1073/pnas.182412299

54. Zhong L, Li B, Mah CS, Govindasamy L, Agbandje-McKenna M, Cooper M, Herzog RW, Zolotukhin I, Warrington KH, Weigel-Van Aken KA, Hobbs JA, Zolotukhin S, Muzyczka N, Srivastava A (2008) Next generation of adeno-associated virus 2 vectors: point mutations in tyrosines lead to high-efficiency transduction at lower doses. Proc Natl Acad Sci U S A 105:7827–7832. doi:10.1073/pnas.0802866105

55. Petrs-Silva H, Dinculescu A, Li Q, Min S-H, Chiodo V, Pang J-J, Zhong L, Zolotukhin S, Srivastava A, Lewin AS, Hauswirth WW (2009) High-efficiency transduction of the mouse retina by tyrosine-mutant AAV serotype vectors. Mol Ther 17:463–471. doi:10.1038/mt.2008.269

56. Markusic DM, Herzog RW, Aslanidi GV, Hoffman BE, Li B, Li M, Jayandharan GR, Ling C, Zolotukhin I, Ma W, Zolotukhin S, Srivastava A, Zhong L (2010) High-efficiency transduction and correction of murine hemophilia B using AAV2 vectors devoid of multiple surface-exposed tyrosines. Mol Ther 18:2048–2056. doi:10.1038/mt.2010.172

57. Martino AT, Basner-Tschakarjan E, Markusic DM, Finn JD, Hinderer C, Zhou S, Ostrov

DA, Srivastava A, Ertl HCJ, Terhorst C, High KA, Mingozzi F, Herzog RW (2013) Engineered AAV vector minimizes in vivo targeting of transduced hepatocytes by capsid-specific CD8+ T cells. Blood 121:2224–2233. doi:10.1182/blood-2012-10-460733

58. Zolotukhin I, Luo D, Gorbatyuk O, Hoffman B, Warrington K, Herzog R, Harrison J, Cao O (2013) Improved adeno-associated viral gene transfer to murine glioma. J Genet Syndr Gene Ther. doi:10.4172/2157-7412.1000133

59. Jang J-H, Koerber JT, Kim J-S, Asuri P, Vazin T, Bartel M, Keung A, Kwon I, Park KI, Schaffer DV (2011) An evolved adeno-associated viral variant enhances gene delivery and gene targeting in neural stem cells. Mol Ther 19:667–675. doi:10.1038/mt.2010.287

60. Kim J-S, Chu HS, Park KI, Won J-I, Jang J-H (2012) Elastin-like polypeptide matrices for enhancing adeno-associated virus-mediated gene delivery to human neural stem cells. Gene Ther 19:329–337. doi:10.1038/gt.2011.84

61. Kotterman MA, Vazin T, Schaffer DV (2015) Enhanced selective gene delivery to neural stem cells in vivo by an adeno-associated viral variant. Development 142:1885–1892. doi:10.1242/dev.115253

62. Chai X, Kong W, Liu L, Yu W, Zhang Z, Sun Y (2014) A viral vector expressing hypoxia-inducible factor 1 alpha inhibits hippocampal neuronal apoptosis. Neural Regen Res 9:1145–1153. doi:10.4103/1673-5374.135317

63. Duan D, Yue Y, Yan Z, Engelhardt JF (2000) A new dual-vector approach to enhance recombinant adeno-associated virus-mediated gene expression through intermolecular cis activation. Nat Med 6:595–598. doi:10.1038/75080

64. Nakai H, Storm TA, Kay MA (2000) Increasing the size of rAAV-mediated expression cassettes in vivo by intermolecular joining of two complementary vectors. Nat Biotechnol 18:527–532. doi:10.1038/75390

65. Sun L, Li J, Xiao X (2000) Overcoming adeno-associated virus vector size limitation through viral DNA heterodimerization. Nat Med 6:599–602. doi:10.1038/75087

66. Gompf HS, Budygin EA, Fuller PM, Bass CE (2015) Targeted genetic manipulations of neuronal subtypes using promoter-specific combinatorial AAVs in wild-type animals. Front Behav Neurosci 9:152. doi:10.3389/fnbeh.2015.00152

67. Denning W, Das S, Guo S, Xu J, Kappes JC, Hel Z (2013) Optimization of the transductional efficiency of lentiviral vectors: effect of sera and polycations. Mol Biotechnol 53:308–314. doi:10.1007/s12033-012-9528-5

68. Naldini L, Blömer U, Gallay P, Ory D, Mulligan R, Gage FH, Verma IM, Trono D (1996) In vivo gene delivery and stable transduction of nondividing cells by a lentiviral vector. Science 272:263–267

69. Dull T, Zufferey R, Kelly M, Mandel RJ, Nguyen M, Trono D, Naldini L (1998) A third-generation lentivirus vector with a conditional packaging system. J Virol 72: 8463–8471

70. Li M, Husic N, Lin Y, Snider BJ (2012) Production of lentiviral vectors for transducing cells from the central nervous system. J Vis Exp:e4031. doi: 10.3791/4031

71. Liu K-C, Lin B-S, Gao A-D, Ma H-Y, Zhao M, Zhang R, Yan H-H, Yi X-F, Lin S-J, Que J-W, Lan X-P (2014) Integrase-deficient lentivirus: opportunities and challenges for human gene therapy. Curr Gene Ther 14:352–364

72. Verhoeyen E, Cosset F-L (2004) Surface-engineering of lentiviral vectors. J Gene Med 6(Suppl 1):S83–S94. doi:10.1002/jgm.494

73. Lei Y, Joo K-I, Wang P (2009) Engineering fusogenic molecules to achieve targeted transduction of enveloped lentiviral vectors. J Biol Eng 3:8. doi:10.1186/1754-1611-3-8

74. Zhang X-Y, Kutner RH, Bialkowska A, Marino MP, Klimstra WB, Reiser J (2010) Cell-specific targeting of lentiviral vectors mediated by fusion proteins derived from sindbis virus, vesicular stomatitis virus, or avian sarcoma/leukosis virus. Retrovirology 7:3. doi:10.1186/1742-4690-7-3

75. Lei Y, Joo K-I, Zarzar J, Wong C, Wang P (2010) Targeting lentiviral vector to specific cell types through surface displayed single chain antibody and fusogenic molecule. Virol J 7:35. doi:10.1186/1743-422X-7-35

76. Kato S, Kobayashi K, Kobayashi K (2014) Improved transduction efficiency of a lentiviral vector for neuron-specific retrograde gene transfer by optimizing the junction of fusion envelope glycoprotein. J Neurosci Methods 227:151–158. doi:10.1016/j.jneumeth.2014.02.015

77. Eleftheriadou I, Trabalza A, Ellison S, Gharun K, Mazarakis N (2014) Specific retrograde transduction of spinal motor neurons using lentiviral vectors targeted to presynaptic NMJ receptors. Mol Ther 22:1285–1298. doi:10.1038/mt.2014.49

78. Yang P (2012) Lentiviral vector mediates exogenous gene expression in adult rat DRG following peripheral nerve remote delivery. J Mol Neurosci 47:173–179. doi:10.1007/s12031-012-9710-z

79. Palfi S, Gurruchaga JM, Ralph GS, Lepetit H, Lavisse S, Buttery PC, Watts C, Miskin J, Kelleher M, Deeley S, Iwamuro H, Lefaucheur

JP, Thiriez C, Fenelon G, Lucas C, Brugières P, Gabriel I, Abhay K, Drouot X, Tani N, Kas A, Ghaleh B, Le Corvoisier P, Dolphin P, Breen DP, Mason S, Guzman NV, Mazarakis ND, Radcliffe PA, Harrop R, Kingsman SM, Rascol O, Naylor S, Barker RA, Hantraye P, Remy P, Cesaro P, Mitrophanous KA (2014) Long-term safety and tolerability of ProSavin, a lentiviral vector-based gene therapy for Parkinson's disease: a dose escalation, open-label, phase 1/2 trial. Lancet 383:1138–1146. doi:10.1016/S0140-6736(13)61939-X

80. Simonato M, Manservigi R, Marconi P, Glorioso J (2000) Gene transfer into neurones for the molecular analysis of behaviour: focus on herpes simplex vectors. Trends Neurosci 23:183–190

81. Epstein AL (2009) HSV-1-derived amplicon vectors: recent technological improvements and remaining difficulties--a review. Mem Inst Oswaldo Cruz 104:399–410

82. Rath P, Shi H, Maruniak JA, Litofsky NS, Maria BL, Kirk MD (2009) Stem cells as vectors to deliver HSV/tk gene therapy for malignant gliomas. Curr Stem Cell Res Ther 4:44–49

83. Friedman AK, Han M-H (2015) The use of herpes simplex virus in ex vivo slice culture. Curr Protoc Neurosci 72:4.36.1–4.36.7. doi:10.1002/0471142301.ns0436s72

84. Mertz KD, Weisheit G, Schilling K, Lüers GH (2002) Electroporation of primary neural cultures: a simple method for directed gene transfer in vitro. Histochem Cell Biol 118:501–506. doi:10.1007/s00418-002-0473-4

85. Inoue T, Krumlauf R (2001) An impulse to the brain—using in vivo electroporation. Nat Neurosci 4:1156–1158

86. Karra D, Dahm R (2010) Transfection techniques for neuronal cells. J Neurosci 30:6171–6177. doi:10.1523/JNEUROSCI.0183-10.2010

87. Kitamura K, Judkewitz B, Kano M, Denk W, Häusser M (2008) Targeted patch-clamp recordings and single-cell electroporation of unlabeled neurons in vivo. Nat Methods 5:61–67. doi:10.1038/nmeth1150

88. Tanaka M, Yanagawa Y, Hirashima N (2009) Transfer of small interfering RNA by single-cell electroporation in cerebellar cell cultures. J Neurosci Methods 178:80–86. doi:10.1016/j.jneumeth.2008.11.025

89. Judkewitz B, Rizzi M, Kitamura K, Häusser M (2009) Targeted single-cell electroporation of mammalian neurons in vivo. Nat Protoc 4:862–869. doi:10.1038/nprot.2009.56

90. Kabakov AY, Lisman JE (2015) Catalytically dead αCaMKII K42M mutant acts as a dominant negative in the control of synaptic strength. PLoS One 10:e0123718. doi:10.1371/journal.pone.0123718

91. Pagès S, Cane M, Randall J, Capello L, Holtmaat A (2015) Single cell electroporation for longitudinal imaging of synaptic structure and function in the adult mouse neocortex in vivo. Front Neuroanat 9:36. doi:10.3389/fnana.2015.00036

92. Leclere PG, Panjwani A, Docherty R, Berry M, Pizzey J, Tonge DA (2005) Effective gene delivery to adult neurons by a modified form of electroporation. J Neurosci Methods 142:137–143. doi:10.1016/j.jneumeth.2004.08.012

93. Gärtner A, Collin L, Lalli G (2006) Nucleofection of primary neurons. Methods Enzymol 406:374–388. doi:10.1016/S0076-6879(06)06027-7

94. Kirton HM, Pettinger L, Gamper N (2013) Transient overexpression of genes in neurons using nucleofection. Methods Mol Biol 998:55–64. doi:10.1007/978-1-62703-351-0_4

95. Kittler JT, Thomas P, Tretter V, Bogdanov YD, Haucke V, Smart TG, Moss SJ (2004) Huntingtin-associated protein 1 regulates inhibitory synaptic transmission by modulating gamma-aminobutyric acid type A receptor membrane trafficking. Proc Natl Acad Sci U S A 101:12736–12741. doi:10.1073/pnas.0401860101

96. Han Q-J, Gao N-N, Guo-QiangMa ZZ-N, Yu W-H, Pan J, Wang Q, Zhang X, Bao L (2013) IPP5 inhibits neurite growth in primary sensory neurons by maintaining TGF-β/Smad signaling. J Cell Sci 126:542–553. doi:10.1242/jcs.114280

97. Bertram B, Wiese S, von Holst A (2012) High-efficiency transfection and survival rates of embryonic and adult mouse neural stem cells achieved by electroporation. J Neurosci Methods 209:420–427. doi:10.1016/j.jneumeth.2012.06.024

98. McCall J, Nicholson L, Weidner N, Blesch A (2012) Optimization of adult sensory neuron electroporation to study mechanisms of neurite growth. Front Mol Neurosci 5:11. doi:10.3389/fnmol.2012.00011

99. Zeitelhofer M, Vessey JP, Xie Y, Tübing F, Thomas S, Kiebler M, Dahm R (2007) High-efficiency transfection of mammalian neurons via nucleofection. Nat Protoc 2:1692–1704. doi:10.1038/nprot.2007.226

100. Hutson TH, Buchser WJ, Bixby JL, Lemmon VP, Moon LDF (2011) Optimization of a 96-well electroporation assay for postnatal rat CNS neurons suitable for cost-effective medium-throughput screening of genes that promote neurite outgrowth. Front Mol Neurosci 4:55. doi:10.3389/fnmol.2011.00055

101. Barry G, Briggs JA, Vanichkina DP, Poth EM, Beveridge NJ, Ratnu VS, Nayler SP, Nones K, Hu J, Bredy TW, Nakagawa S, Rigo F, Taft RJ, Cairns MJ, Blackshaw S, Wolvetang EJ, Mattick JS (2013) The long non-coding RNA Gomafu is acutely regulated in response to neuronal activation and involved in schizophrenia-associated alternative splicing. Mol Psychiatry 19:486–494. doi:10.1038/mp.2013.45

102. Chen Y, Wang B, Liu D, Li JJ, Xue Y, Sakata K, Zhu L, Heldt SA, Xu H, Liao F-F (2014) Hsp90 chaperone inhibitor 17-AAG attenuates Aβ-induced synaptic toxicity and memory impairment. J Neurosci 34:2464–2470. doi:10.1523/JNEUROSCI.0151-13.2014

103. Scherer F, Anton M, Schillinger U, Henke J, Bergemann C, Krüger A, Gänsbacher B, Plank C (2002) Magnetofection: enhancing and targeting gene delivery by magnetic force in vitro and in vivo. Gene Ther 9:102–109. doi:10.1038/sj.gt.3301624

104. Plank C, Schillinger U, Scherer F, Bergemann C, Rémy J-S, Krötz F, Anton M, Lausier J, Rosenecker J (2003) The magnetofection method: using magnetic force to enhance gene delivery. Biol Chem 384:737–747. doi:10.1515/BC.2003.082

105. Plank C, Zelphati O, Mykhaylyk O (2011) Magnetically enhanced nucleic acid delivery. Ten years of magnetofection-progress and prospects. Adv Drug Deliv Rev 63:1300–1331. doi:10.1016/j.addr.2011.08.002

106. Underhill SM, Wheeler DS, Li M, Watts SD, Ingram SL, Amara SG (2014) Amphetamine modulates excitatory neurotransmission through endocytosis of the glutamate transporter EAAT3 in dopamine neurons. Neuron 83:404–416. doi:10.1016/j.neuron.2014.05.043

107. Buerli T, Pellegrino C, Baer K, Lardi-Studler B, Chudotvorova I, Fritschy J-M, Medina I, Fuhrer C (2007) Efficient transfection of DNA or shRNA vectors into neurons using magnetofection. Nat Protoc 2:3090–3101. doi:10.1038/nprot.2007.445

108. Alavian KN, Li H, Collis L, Bonanni L, Zeng L, Sacchetti S, Lazrove E, Nabili P, Flaherty B, Graham M, Chen Y, Messerli SM, Mariggio MA, Rahner C, McNay E, Shore GC, Smith PJS, Hardwick JM, Jonas EA (2011) Bcl-xL regulates metabolic efficiency of neurons through interaction with the mitochondrial F1FO ATP synthase. Nat Cell Biol 13:1224–1233. doi:10.1038/ncb2330

109. Charrier C, Joshi K, Coutinho-Budd J, Kim J-E, Lambert N, de Marchena J, Jin W-L, Vanderhaeghen P, Ghosh A, Sassa T, Polleux F (2012) Inhibition of SRGAP2 function by its human-specific paralogs induces neoteny during spine maturation. Cell 149:923–935. doi:10.1016/j.cell.2012.03.034

110. Mairet-Coello G, Courchet J, Pieraut S, Courchet V, Maximov A, Polleux F (2013) The CAMKK2-AMPK kinase pathway mediates the synaptotoxic effects of Aβ oligomers through tau phosphorylation. Neuron 78:94–108. doi:10.1016/j.neuron.2013.02.003

111. Terenzio M, Golding M, Russell MRG, Wicher KB, Rosewell I, Spencer-Dene B, Ish-Horowicz D, Schiavo G (2014) Bicaudal-D1 regulates the intracellular sorting and signalling of neurotrophin receptors. EMBO J 33:1582–1598. doi:10.15252/embj.201387579

112. Wang R, Palavicini JP, Wang H, Maiti P, Bianchi E, Xu S, Lloyd BN, Dawson-Scully K, Kang DE, Lakshmana MK (2014) RanBP9 overexpression accelerates loss of dendritic spines in a mouse model of Alzheimer's disease. Neurobiol Dis 69:169–179. doi:10.1016/j.nbd.2014.05.029

113. Fallini C, Bassell GJ, Rossoll W (2010) High-efficiency transfection of cultured primary motor neurons to study protein localization, trafficking, and function. Mol Neurodegener 5:17. doi:10.1186/1750-1326-5-17

114. Fallini C, Rouanet JP, Donlin-Asp PG, Guo P, Zhang H, Singer RH, Rossoll W, Bassell GJ (2014) Dynamics of survival of motor neuron (SMN) protein interaction with the mRNA-binding protein IMP1 facilitates its trafficking into motor neuron axons. Dev Neurobiol 74:319–332. doi:10.1002/dneu.22111

115. Fallini C, Bassell GJ, Rossoll W (2012) The ALS disease protein TDP-43 is actively transported in motor neuron axons and regulates axon outgrowth. Hum Mol Genet 21:3703–3718. doi:10.1093/hmg/dds205

116. Sapet C, Laurent N, de Chevigny A, le Gourrierec L, Bertosio E, Zelphati O, Béclin C (2011) High transfection efficiency of neural stem cells with magnetofection. Biotechniques 50:187–189. doi:10.2144/000113628

117. Pickard MR, Adams CF, Barraud P, Chari DM (2015) Using magnetic nanoparticles for gene transfer to neural stem cells: stem cell propagation method influences outcomes. J Funct Biomater 6:259–276. doi:10.3390/jfb6020259

118. Restituito S, Couve A, Bawagan H, Jourdain S, Pangalos MN, Calver AR, Freeman KB, Moss SJ (2005) Multiple motifs regulate the trafficking of GABA(B) receptors at distinct checkpoints within the secretory pathway. Mol Cell Neurosci 28:747–756. doi:10.1016/j.mcn.2004.12.006

119. Bounhar Y, Zhang Y, Goodyer CG, LeBlanc A (2001) Prion protein protects human neurons against Bax-mediated apoptosis. J Biol Chem 276:39145–39149. doi:10.1074/jbc.C100443200

120. Roucou X, Giannopoulos PN, Zhang Y, Jodoin J, Goodyer CG, LeBlanc A (2005) Cellular prion protein inhibits proapoptotic Bax conformational change in human neurons and in breast carcinoma MCF-7 cells. Cell Death Differ 12:783–795. doi:10.1038/sj.cdd.4401629

121. Li Z, Gu X, Sun L, Wu S, Liang L, Cao J, Lutz BM, Bekker A, Zhang W, Tao Y-X (2015) Dorsal root ganglion myeloid zinc finger protein 1 contributes to neuropathic pain after peripheral nerve trauma. Pain 156:711–721. doi:10.1097/j.pain.0000000000000103

122. Tsutajima J, Kunitake T, Wakazono Y, Takamiya K (2013) Selective injection system into hippocampus CA1 via monitored theta oscillation. PLoS One 8:e83129. doi:10.1371/journal.pone.0083129

123. Kole AJ, Knight ERW, Deshmukh M (2011) Activation of apoptosis by cytoplasmic microinjection of cytochrome c. J Vis Exp. doi:10.3791/2773

124. Kohara K, Kitamura A, Morishima M, Tsumoto T (2001) Activity-dependent transfer of brain-derived neurotrophic factor to postsynaptic neurons. Science 291:2419–2423. doi:10.1126/science.1057415

125. Kittler JT, Wang J, Connolly CN, Vicini S, Smart TG, Moss SJ (2000) Analysis of GABAA receptor assembly in mammalian cell lines and hippocampal neurons using gamma 2 subunit green fluorescent protein chimeras. Mol Cell Neurosci 16:440–452. doi:10.1006/mcne.2000.0882

126. Zhang Y, Yu L-C (2008) Single-cell microinjection technology in cell biology. Bioessays 30:606–610. doi:10.1002/bies.20759

127. Lo DC (2001) Neuronal transfection using particle-mediated gene transfer. Curr Protoc Neurosci Chapter 3:Unit 3.15. doi:10.1002/0471142301.ns0315s05

128. Dib-Hajj SD, Choi JS, Macala LJ, Tyrrell L, Black JA, Cummins TR, Waxman SG (2009) Transfection of rat or mouse neurons by biolistics or electroporation. Nat Protoc 4:1118–1126. doi:10.1038/nprot.2009.90

129. Klein RM, Wolf ED, Wu R, Sanford JC (1992) High-velocity microprojectiles for delivering nucleic acids into living cells. 1987. Biotechnology 24:384–386

130. Klein TM, Fromm M, Weissinger A, Tomes D, Schaaf S, Sletten M, Sanford JC (1988) Transfer of foreign genes into intact maize cells with high-velocity microprojectiles. Proc Natl Acad Sci U S A 85:4305–4309

131. Arsenault J, Nagy A, Henderson JT, O'Brien JA (2014) Regioselective biolistic targeting in organotypic brain slices using a modified gene gun. J Vis Exp:e52148. doi: 10.3791/52148

132. Gamper N, Shapiro MS (2006) Exogenous expression of proteins in neurons using the biolistic particle delivery system. Methods Mol Biol 337:27–38. doi:10.1385/1-59745-095-2:27

133. Carta M, Opazo P, Veran J, Athané A, Choquet D, Coussen F, Mulle C (2013) CaMKII-dependent phosphorylation of GluK5 mediates plasticity of kainate receptors. EMBO J 32:496–510. doi:10.1038/emboj.2012.334

134. Oliveira AF, Yasuda R (2014) Neurofibromin is the major ras inactivator in dendritic spines. J Neurosci 34:776–783. doi:10.1523/JNEUROSCI.3096-13.2014

135. Lachance-Touchette P, Choudhury M, Stoica A, Di Cristo G, Cossette P (2014) Single-cell genetic expression of mutant GABAA receptors causing human genetic epilepsy alters dendritic spine and GABAergic bouton formation in a mutation-specific manner. Front Cell Neurosci 8:317. doi:10.3389/fncel.2014.00317

136. O'Brien JA, Holt M, Whiteside G, Lummis SC, Hastings MH (2001) Modifications to the hand-held gene gun: improvements for in vitro biolistic transfection of organotypic neuronal tissue. J Neurosci Methods 112:57–64

137. O'Brien JA, Lummis SCR (2006) Biolistic transfection of neuronal cultures using a hand-held gene gun. Nat Protoc 1:977–981. doi:10.1038/nprot.2006.145

138. Arsenault J, O'Brien JA (2013) Optimized heterologous transfection of viable adult organotypic brain slices using an enhanced gene gun. BMC Res Notes 6:544. doi:10.1186/1756-0500-6-544

139. Felgner PL, Gadek TR, Holm M, Roman R, Chan HW, Wenz M, Northrop JP, Ringold GM, Danielsen M (1987) Lipofection: a highly efficient, lipid-mediated DNA-transfection procedure. Proc Natl Acad Sci U S A 84:7413–7417

140. Ohki EC, Tilkins ML, Ciccarone VC, Price PJ (2001) Improving the transfection efficiency of post-mitotic neurons. J Neurosci Methods 112:95–99

141. Dalby B, Cates S, Harris A, Ohki EC, Tilkins ML, Price PJ, Ciccarone VC (2004) Advanced transfection with Lipofectamine 2000 reagent: primary neurons, siRNA, and high-throughput applications. Methods 33:95–103. doi:10.1016/j.ymeth.2003.11.023

142. Guo Z, Cao Y-Q (2014) Over-expression of TRESK K(+) channels reduces the excitability of trigeminal ganglion nociceptors. PLoS One 9:e87029. doi:10.1371/journal.pone.0087029

143. Williams DJ, Puhl HL, Ikeda SR (2010) A simple, highly efficient method for heterologous expression in mammalian primary neurons using cationic lipid-mediated mRNA transfection. Front Neurosci 4:181. doi:10.3389/fnins.2010.00181

144. McLenachan S, Zhang D, Palomo ABA, Edel MJ, Chen FK (2013) mRNA transfection of mouse and human neural stem cell cultures. PLoS One 8:e83596. doi:10.1371/journal.pone.0083596

145. Wiesenhofer B, Kaufmann WA, Humpel C (1999) Improved lipid-mediated gene transfer in C6 glioma cells and primary glial cells using FuGene™. J Neurosci Methods 92:145–152. doi:10.1016/S0165-0270(99)00108-9

146. Wiesenhofer B, Humpel C (2000) Lipid-mediated gene transfer into primary neurons using FuGene: comparison to C6 glioma cells and primary glia. Exp Neurol 164:38–44. doi:10.1006/exnr.2000.7414

147. Tinsley RB, Faijerson J, Eriksson PS (2006) Efficient non-viral transfection of adult neural stem/progenitor cells, without affecting viability, proliferation or differentiation. J Gene Med 8:72–81. doi:10.1002/jgm.823

148. Tönges L, Lingor P, Egle R, Dietz GPH, Fahr A, Bähr M (2006) Stearylated octaarginine and artificial virus-like particles for transfection of siRNA into primary rat neurons. RNA 12:1431–1438. doi:10.1261/rna.2252206

149. Butcher AJ, Torrecilla I, Young KW, Kong KC, Mistry SC, Bottrill AR, Tobin AB (2009) N-methyl-D-aspartate receptors mediate the phosphorylation and desensitization of muscarinic receptors in cerebellar granule neurons. J Biol Chem 284:17147–17156. doi:10.1074/jbc.M901031200

150. Dai S-H, Chen T, Wang Y-H, Zhu J, Luo P, Rao W, Yang Y-F, Fei Z, Jiang X-F (2014) Sirt3 protects cortical neurons against oxidative stress via regulating mitochondrial Ca2+ and mitochondrial biogenesis. Int J Mol Sci 15:14591–14609. doi:10.3390/ijms150814591

151. Köhrmann M, Haubensak W, Hemraj I, Kaether C, Leßmann VJ, Kiebler MA (1999) Fast, convenient, and effective method to transiently transfect primary hippocampal neurons. J Neurosci Res 58:831–835. doi:10.1002/(SICI)1097-4547(19991215)58:6<831::AID-JNR10>3.0.CO;2-M

152. Goetze B, Grunewald B, Baldassa S, Kiebler M (2004) Chemically controlled formation of a DNA/calcium phosphate coprecipitate: application for transfection of mature hippocampal neurons. J Neurobiol 60:517–525. doi:10.1002/neu.20073

153. Dudek H, Ghosh A, Greenberg ME (2001) Calcium phosphate transfection of DNA into neurons in primary culture. Curr Protoc Neurosci Chapter 3:Unit 3.11. doi:10.1002/0471142301.ns0311s03

154. Dahm R, Zeitelhofer M, Götze B, Kiebler MA, Macchi P (2008) Visualizing mRNA localization and local protein translation in neurons. Methods Cell Biol 85:293–327. doi:10.1016/S0091-679X(08)85013-3

155. Jiang M, Chen G (2006) High Ca2+–phosphate transfection efficiency in low-density neuronal cultures. Nat Protoc 1:695–700. doi:10.1038/nprot.2006.86

156. Sun M, Bernard LP, Dibona VL, Wu Q, Zhang H (2013) Calcium phosphate transfection of primary hippocampal neurons. J Vis Exp:e50808. doi:10.3791/50808

157. Junquera E, Aicart E (2015) Recent progress in gene therapy to deliver nucleic acids with multivalent cationic vectors. Adv Colloid Interface Sci 233:161–175. doi:10.1016/j.cis.2015.07.003

158. Becattini G, Mattos LS, Caldwell DG (2014) A fully automated system for adherent cells microinjection. IEEE J Biomed Heal Inform 18:83–93. doi:10.1109/JBHI.2013.2248161

Chapter 5

Reprogramming of Mouse Fibroblasts to Induced Oligodendrocyte Progenitor Cells

Robert T. Karl, Angela M. Lager, Fadi J. Najm, and Paul J. Tesar

Abstract

Oligodendrocyte progenitor cells are the major myelinating cell type of the central nervous system and their dysfunction contributes to a variety of neurological diseases. However, direct access to oligodendrocyte progenitor cells has been challenging. Recently, cellular reprogramming technologies have demonstrated the ability to directly convert one cell type to another. This chapter describes the methods for the generation of induced oligodendrocyte progenitor cells from mouse embryonic fibroblasts by overexpression of defined transcription factors. We also describe pertinent assays used to confirm transgene expression, reprogrammed cell identity, and terminal differentiation to mature oligodendrocytes.

Key words Reprogramming, Immunostaining, Quantitative PCR, RNA-seq, Oligodendrocyte progenitor cells, Lentivirus, Mouse embryonic fibroblasts

1 Introduction

Diseases caused by loss or dysfunction of myelin in the central nervous system (CNS) affect millions of people worldwide. These diseases often lead to severe disability and include childhood congenital leukodystrophies as well as adult onset disorders like multiple sclerosis. Oligodendrocyte progenitor cells (OPCs) are the major source of myelinating oligodendrocytes in the CNS and cell transplantation studies have shown the potential to ameliorate disease symptoms [1–4]. Progress in the field has been hindered by lack of access to autologous populations of OPCs [5]. Cells generated using direct lineage reprogramming may represent a potential avenue for generating large quantities of OPCs to be used for cell-based transplantation or molecular studies.

Recently, we have shown the ability to efficiently derive large, pure populations of OPCs capable of differentiating into mature myelinating oligodendrocytes from mouse pluripotent stem cells using defined developmental pathways and signals [6]. We sought to leverage our knowledge of the oligodendrocyte lineage development

Amit K. Srivastava et al. (eds.), *Stem Cell Technologies in Neuroscience*, Neuromethods, vol. 126,
DOI 10.1007/978-1-4939-7024-7_5, © Springer Science+Business Media LLC 2017

to directly convert mouse fibroblasts to induced OPCs (iOPCs) [7]. This is achieved by inducing the expression of a subset of exogenous lineage transcription factors. These iOPCs exhibit morphological properties of bona fide OPCs, globally express OPC genes, and are able to generate compact myelin. This chapter provides detailed instructions to generate these cells.

2   Materials

2.1   Generating iOPCs

2.1.1   Culturing MEFs

1. Embryonic day 13.5 (E13.5) isolated MEFs [8] (*see* **Notes 1** and **2**).

2. Nunc culture plates with the Nunclon Δ-treated surface (6-well) (Nunc; 40675) (*see* **Note 3**).

3. MEF culture medium: Dulbecco's modified Eagle's medium (Invitrogen; 11960) supplemented with 10% fetal bovine serum (FBS) (Invitrogen; 16000044), 2 mM GlutaMAX (Invitrogen; 35050061), 1× nonessential amino acids (Invitrogen; 11140-50), and 0.1 mM 2-mercaptoethanol (Sigma; M3148). Media components are combined and filtered through a 0.22 μm polyethersulfone filtration unit (Millipore; SGCPU05RE) and stored at 4 °C for up to 1 month.

4. 0.25% trypsin–EDTA for passaging (Invitrogen; 25200).

5. Phosphate-buffered saline (PBS) without calcium and magnesium for passaging.

2.1.2   Lentivirus Production

1. HEK293T cell line (*see* **Note 4**).

2. Rat Tail Collagen I coated T75 flasks (BD; 356485).

3. MEF culture medium.

4. Round bottom polystyrene tubes (15 mL).

5. 0.45 μm polyvinylidene difluoride membrane filtration unit (Millipore; SE1M003M00).

6. Mini vortex mixer.

7. pLVX Tet-On viral plasmids of *Nkx6.2*, *Olig2*, *Sox10*, *Olig1*, *Nkx2.2*, *Myt1*, *St18*, and *Myrf* (*see* **Notes 5** and **6**).

8. Lenti-X HTX Packaging Mix and Lenti-Phos or Cal-Phos Mammalian Transfection Kit (Clontech; 631247 & 631312).

9. Bleach.

2.1.3   Infecting MEFs

1. Nunc culture plates with the Nunclon Δ-treated surface (6-well).

2. MEF culture medium.

3. 1000× Polybrene solution (8 mg/mL, Sigma; H9268) for viral infection.

4. 500× Doxycycline solution (1 mg/mL, Clonetech; 631311) for virus induction.

5. Bleach.

2.2 Culturing and Maintaining iOPCs

2.2.1 Passaging and Culturing iOPCs

1. Nunclon Δ-treated cell culture plates (6-well and 12-well) pretreated with poly-L-ornithine (Sigma; 3655) and laminin (Sigma; L2020) (*see* **Note 3**).

2. TrypLE Select for passaging (Invitrogen; 12563).

3. OPC medium: Dulbecco's Modified Eagle's Medium/F12 supplemented (Invitrogen; 11320) with 1× N-2 Max (R&D Systems; AR009), 1× B-27 without vitamin A (Invitrogen; 12587), and 2 mM GlutaMAX. Media components are combined and filtered through a 0.22 μm polyethersulfone filtration unit and stored at 4 °C for up to 1 month in the dark.

4. OPC growth medium: OPC medium supplemented with SHH (200 ng/mL, R&D Systems; 1845-SH), FGF2 (20 ng/mL, R&D Systems; 233-FB), and PDGF-AA (20 ng/mL, R&D Systems; 221-AA) prior to use. SHH, FGF, and PDGF-AA are dissolved in 0.1% bovine serum albumin fraction V in PBS to a concentration of 200 μg/mL (SHH) and 10 μg/mL (FGF2 and PDGF-AA). Aliquots of growth factor and recombinant protein are stored at −80 °C, and used within 6 months. Once thawed, aliquots are stored at 4 °C and used within 2 weeks.

5. 500× Doxycycline solution (1 mg/mL) for virus induction

2.2.2 Freezing iOPCs

1. 1.8 mL cryovials (Thermo; 375418).

2. Cryo-freezing container: Nalgene "Mr. Frosty." Container filled with 250 mL of isopropanol and changed after four uses.

3. Freezing medium: 60% Dulbecco's Modified Eagle's Medium, 20% FBS, 20% DMSO. Media components are combined in the above order and filtered through a 0.22 μm syringe filter. Store media at 4 °C and used for up to 1 week.

2.2.3 Thawing iOPCs

1. OPC medium warmed in a 37 °C water bath.

2. OPC growth medium warmed in a 37 °C water bath.

3. Nunc culture plates with the Nunclon Δ-treated surface (6-well) pretreated with poly-L-ornithine and laminin.

2.3 Characterization of iOPCs

2.3.1 Quantitative PCR

1. TRIzol (Invitrogen; 15596026).

2. Chloroform.

3. 2.0 mL Phase Lock Gel Heavy (5 Prime; 2302830).

4. miRNeasy mini Kit (Qiagen; 217004).

5. NanoDrop.

6. SuperScript III Reverse Transcriptase (Invitrogen; 18080093).

7. TaqMan Gene Expression Master Mix (Invitrogen; 4369016).

8. TaqMan probes: *Nkx6.2* (Mm00807812_g1), *Olig2* (AJVI3GC, custom), *Sox10* (Mm01300162_m1), *Olig1* (Mm00497537_s1), *Nkx2.2* (Mm01275962_m1), *Myt1* (Mm00456190_m1), *St18* (Mm01236999_m1), and *Myrf* (Mm01194959_m1) *Gapdh* (Mm99999915_g1). For detection of endogenous *Olig2* (Mm01210556_m1) (*see* **Note 7**).

9. 7300 Real-Time PCR System (Applied Biosystems) or equivalent Real-Time PCR system.

2.3.2 Flow Cytometry Analysis

1. Flow cytometry solution: Supplement OPC medium with 2 mM EDTA and 0.5% bovine serum albumin Fraction V and filter through a 0.22 μm filter. Store solution at 4 °C and use for up to 10 days.

2. Antibodies: APC-conjugated A2B5 (Miltenyi Biotec; 130-093-582; 1:11) and conjugated isotype control (Miltenyi Biotec; 130-093-176; 1:11).

3. 5 mL round bottom polystyrene tube with 0.35 μm cell strainer snap cap (Falcon; 352235).

4. 1.5 mL microcentrifuge tubes.

2.3.3 Differentiation to Oligodendrocytes

1. Nunclon Δ-treated cell culture plates (6-well) pretreated with poly-L-ornithine and laminin.

2. TrypLE Select for passaging.

3. OPC medium for washes.

4. OPC differentiation medium: OPC medium supplemented with SHH (200 ng/mL), Noggin (100 ng/mL, R&D Systems 3344-NG), cAMP (50 μM, D0260), IGF (100 ng/mL, R&D Systems 291-G1), NT-3 (10 ng/mL, R&D Systems 267-N3), and T3 (40 ng/mL, Sigma T6397) prior to use. SHH, Noggin, IGF, and NT-3 are dissolved in 0.1% bovine serum albumin fraction V in 1× PBS to a concentration of 200 μg/mL (SHH), 100 μg/mL (Noggin and IGF), or 10 μg/mL (NT-3). cAMP is dissolved in sterile water to a concentration of 50 mM. Aliquots of growth factors and recombinant proteins are stored at −80 °C, and used within 6 months. T3 is dissolved in 0.1 N NaOH and 1× PBS to a concentration of 40 μg/mL and stored as aliquots at −20 °C, and used within 1 year. Once thawed, aliquots are stored at 4 °C and used within 2 weeks.

2.3.4 Immunostaining of OPC and Oligodendrocyte Markers

1. 4% paraformaldehyde in PBS: In a fume hood combine 10 mL of 10% paraformaldehyde solution (Electron Microscopy Sciences; 15710), 4 mL 10× PBS, and 26 mL of sterile water. Store paraformaldehyde at 4 °C and use for up to 10 days.

2. 0.2% Triton X-100 in PBS: Supplement 500 mL of 1× PBS solution with 1 mL of Triton X-100 solution (Sigma; T8787).

Stir with a magnetic stir bar at room temperature for up to 1 h. Store Triton X-100 solution at room temperature and use for up to 1 year.

3. Blocking solution: Dilute 1 mL of donkey serum (Abcam; ab7475) with 9 mL of 1× PBS solution and filter through a 0.22 μm syringe filter. Store media at 4 °C and use for up to 10 days.

4. Primary antibodies: Olig2 (Millipore; AB9610; 1:1000), Sox10 (R&D Systems; AF2864; 1:1000), and Nkx6.2 (Abcam; ab58705; 1:500), Mbp (Covance; SMI99; 1:500 or Abcam; ab7349; 1:100).

5. Secondary antibodies: Appropriate Alexa Fluor® secondary antibody (Invitrogen).

6. DAPI stock solution: Add 1 mL of water to 1 mg of DAPI powder (Sigma; D8417). Store DAPI solution at 4 °C and use for up to 1 year.

7. Phosphate-buffered saline (PBS) without calcium and magnesium.

2.3.5 Global Gene Expression Analysis by RNA Sequencing

1. TRIzol.

2. Chloroform.

3. 2.0 mL Phase Lock Gel Heavy.

4. miRNeasy mini Kit (Qiagen; 217004).

5. 70% ethanol.

6. Illumina Truseq RNA Sample Preparation Kit. (Illumina; RS-122-2001).

3   Methods

3.1   Generation of iOPCs

3.1.1   Production of Lentivirus

1. Seed 5–6 × 10⁶ HEK293T cells in eight rat tail collagen I coated T75 flasks with 12 mL of MEF media and incubate at 37 °C overnight.

2. Aspirate MEF media from each of the T75 flasks and add 12 mL of fresh pre-warmed MEF media and incubate at 37 °C for at least 2 h. Cells should be 60–70% confluent.

3. For each virus: In a polystyrene tube add 15 μL Lenti-X packaging mix, 3 μg pLVX viral plasmid, and sterile water up to 438 μL total volume.

4. Add 62 μL Ca⁺ solution to each of the polystyrene tubes and vortex.

5. While vortexing, add dropwise 500 μL HBSS solution.

6. Incubate the polystyrene tubes at room temperature for 10 min.

7. Gently triturate the transfection solution and add it dropwise to respective HEK293T T75 flask.

8. Incubate the flasks at 37 °C overnight.

9. Aspirate MEF media from each of the T75 flasks and add 10 mL of fresh pre-warmed MEF media and incubate at 37 °C overnight (*see* **Note 8**).

10. Collect lentivirus supernatant into individual 50 mL conical tubes and centrifuge at 610 × g for 15 min at 4 °C.

11. Filter supernatant through a 0.45 μm polyvinylidene difluoride membrane filtration unit.

12. Add 10 mL of pre-warmed MEF media to each T75 flask for collecting viral media the following day.

13. For subsequent lentivirus supernatant collection the following day, collect fresh lentivirus supernatant as described above in Sect. 3.1.1 **steps 10** and **11**. At this point the lentiviral flasks should be discarded by rinsing the flask with a 10% bleach solution.

3.1.2   Lentiviral Infection of MEFs

1. Seed 2 × 10^5 MEFs in a well of a 6-well Nunclon Δ-treated plate with 2 mL of MEF media and incubated at 37 °C overnight (*see* **Note 9**).

2. Supplement fresh collected lentivirus with 8 μg/mL polybrene as described in Sect. 3.1.1 **step 11**.

3. Aspirate media from the MEF 6-wells and add equal volumes (125 μL) of fresh *Sox10*, *Olig2*, *Nkx6.2*, *Olig1*, *Nkx2.2*, *Myt1*, *ST18*, and *Myrf* lentivirus supernatant to each well and incubate at 37 °C for 3 h (*see* **Note 10**).

4. Store remaining lentivirus supernatant at 4 °C.

5. Remove lentivirus supernatant from each MEF 6-well and add equal volumes (125 μL) of fresh *Sox10*, *Olig2*, *Nkx6.2*, *Olig1*, *Nkx2.2*, *Myt1*, *ST18*, and *Myrf* lentivirus supernatant to each well and incubate at 37 °C for 2 h (*see* **Note 10**).

6. Without removing lentivirus add 1 mL of pre-warmed MEF media to each 6-well and incubate at 37 °C overnight.

7. The following day repeat MEF infection as described above in Sect. 3.1.2 **steps 2** through **5**.

8. Remove lentivirus and replace with 2 mL of pre-warmed MEF media supplemented with 2 μg/mL doxycycline and incubate at 37 °C for 36 h. At this point the cells are termed "day 0."

3.2   Culture and Maintenance of iOPCs

3.2.1   Pretreating Nunclon Δ-Treated Cell Culture Plates

1. Prepare 10 mg/mL stock solution of poly-L-ornithine by dissolving 50 mg of poly-L-ornithine in 5 mL of sterile water and filter through a 0.22 μm polyethersulfone filtration unit. Stock solution can be stored at −20 °C in 250 μL aliquots.

2. Prepare 0.1 mg/mL working solution of poly-L-ornithine by diluting a 250 μL aliquot of 10 mg/mL poly-L-ornithine in 24.75 mL of sterile water.

3. Add 2 ml of 0.1 mg/mL poly-L-ornithine per Nunclon Δ-treated 6-well and 1 mL of 0.1 mg/mL poly-L-ornithine per Nunclon Δ-treated 12-well and incubate at 37 °C overnight.

4. Aspirate the poly-L-ornithine, rinse once with 1 mL of sterile water per Nunclon Δ-treated 6-well and 0.5 mL sterile water per Nunclon Δ-treated 12-well, aspirate water, and allow wells to dry in a cell culture hood at room temperature overnight (*see* **Note 11**).

5. Prepare 10 μg/mL working solution of laminin by diluting 20 μL of 1 mg/mL laminin in 2 mL of sterile PBS.

6. Add 1 mL of 10 μg/mL laminin to poly-L-ornithine pretreated Nunclon Δ-treated 6-wells and 0.5 mL to poly-L-ornithine pretreated Nunclon Δ-treated 12-wells and incubate at 37 °C for a minimum of 1 h before seeding iOPCs.

3.2.2 Passaging Infected MEFs

1. On "day 3," 36 h post MEF infection, aspirate media from the 6-wells and rinse with 2 mL 1× PBS.

2. Add 0.5 mL of TrypLE to each 6-well and incubate at 37 °C for 5 min.

3. Dissociate cells by trituration and transfer cell suspension to a 15 mL conical tube.

4. Rinse remaining cells from the 6-well with OPC media and add to the 15 mL conical tube and centrifuge for 4 min at $300 \times g$.

5. Aspirate supernatant and suspend the cell pellet with 1 mL of OPC media using a P1000 pipette followed by the addition of 9 mL of OPC media. Centrifuge for 4 min at $300 \times g$. Repeat once.

6. Aspirate supernatant, suspend the cell pellet with 1 mL of OPC media using a P1000, and determine the number of cells through the use of a hemocytometer or automated cell counter. Add 9 mL of OPC media and centrifuge for 4 min at $300 \times g$.

7. Aspirate supernatant and suspend the cell pellet with OPC media using a P1000 pipette.

8. Aspirate laminin from the culture wells and seed 2×10^5 cells in each Nunclon Δ-treated 6-well pretreated with poly-L-ornithine and laminin with 2 mL of OPC growth media supplemented with 2 μg/mL doxycycline; for immunocytochemistry seed 8×10^4 cells into two Nunclon Δ-treated 12-wells (each well gets 4×10^4 cells) pretreated with poly-L-ornithine and laminin with 1 mL of OPC growth media supplemented with 2 μg/mL doxycycline (*see* **Note 12**).

9. Incubate cells at 37 °C for 24 h (*see* **Note 13**).

10. Freeze down remaining cells as described in Sect. 3.2.4.

3.2.3 Culturing iOPCs

1. Starting on "day 5" OPC culture medium is changed every other day up until "day 14" or "day 21."

2. On "day 14" or "day 21" cells can be passaged for flow cytometry analysis, oligodendrocyte differentiation, cryopreservation, or continued culturing. Morphological differences between reprogrammed iOPCs and starting MEF population can be seen in Fig. 1a and b

3.2.4 Cryopreservation of Infected MEFs and iOPCs

1. Lift and suspend cells as described in Sect. 3.2.2 **steps 1** through 7.

2. In a 1.8 mL cryovial add 5×10^5 cells in a volume of 0.5 mL OPC media.

3. Add 0.5 mL of freezing media to the cryovial.

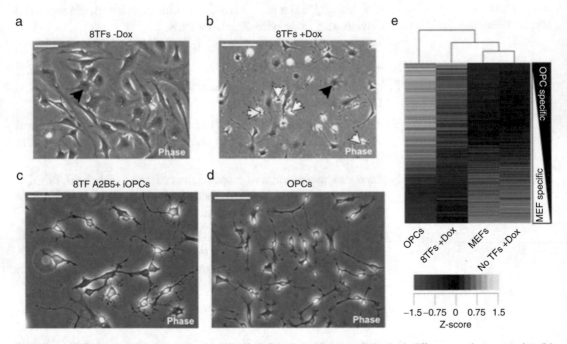

Fig. 1 (**a**, **b**) Phase-contrast images highlighting the dramatic morphological differences between day-21 8TF-uninduced (**a**; −Dox) and induced (**b**; +Dox) MEFs. Nonreprogrammed cells have typical fibroblast morphology (black arrowheads) whereas a portion of the 8TF-induced cultures show the characteristic bipolar morphology of OPCs (white arrowheads). (**c,d**) Phase contrast images highlighting the bipolar morphology of 8TF-induced MEFs (sorted at day 21 for both Plp1-eGFP and A2B5; denoted as 8TF A2B5⁺ iOPCs) (**d**) compared to bona fide OPCs. (**e**) Clustered heat map of z-scored global gene expression values comparing pluripotent stem cell–derived bona fide OPCs, 8TF-induced MEFs (8TFs +Dox), MEFs and uninfected MEFs plus doxycycline (No TFs +Dox). Plot is rank ordered with OPC-specific genes at the top and MEF-specific genes at the bottom and includes >13,000 genes for which there was signal above background in at least one of the samples. Scale bars, 50 μm (**a**, **c**, **d**), 100 μm (**b**)

4. Place tubes in a room-temperature cyro-freezing container and immediately place at −80 °C.

5. The following day transfer vials to a liquid nitrogen freezer for long term storage.

3.2.5 Thawing Infected MEFs and iOPCs

1. Remove cryovial from liquid nitrogen storage and thaw in a 37 °C water bath until only a small ice crystal remains.

2. To nearly thawed cryovial add 1 mL of pre-warmed OPC media dropwise. Transfer cell suspension to 8 mL of pre-warmed OPC media and centrifuge for 4 min at 300 × *g*.

3. Aspirate supernatant and suspend in 10 mL pre-warmed N2B27. Centrifuge for 4 min at 300 × *g*.

4. Aspirate supernatant and suspend the cell pellet in 4 mL of OPC media supplemented with 200 ng/mL SHH, 20 ng/mL FGF2, 20 ng/mL PDGF-AA, and 2 μg/mL doxycycline.

5. Aspirate the laminin from the culture wells and split the cell suspension equally between two Nunclon Δ-treated 6-wells pretreated with poly-L-ornithine and laminin (*see* **Note 14**).

6. Change OPC culture media every other day.

3.3 Characterization of iOPCs
3.3.1 Quantitative PCR (See Note 15)

1. Isolate and lyse cells in 1 mL of TRIzol (*see* **Note 16**).

2. Homogenize by pipetting or vortexing (*see* **Note 17**).

3. Allow samples to incubate for 5 min at room temperature.

4. Add 200 μL chloroform to each sample. Shake vigorously for 15 s.

5. Transfer sample to Phase-Lock tubes (*see* **Note 18**).

6. Incubate at room temperature for 2–3 min.

7. Spin at 12,000 × *g* at 4 °C for 15 min (*see* **Note 19**).

8. From this point forward continue isolation of RNA from **Step 8** of the miRNeasy handbook.

9. Check the concentration of RNA directly following purification using NanoDrop spectrophotometer. Store samples at −80 °C if not using immediately.

10. For each sample add 400 ng of total RNA to a PCR tube.

11. To each sample add: 0.1 μL Random Primers, 1 μL 10 mM dNTPs, and dilute with water to a total volume of 13 μL. Incubate at 65 °C for 5 min (*see* **Note 20**).

12. After the incubation, to each sample add: 4 μL 5× First Strand Buffer, 1 μL 0.1 M DTT, 1 μL RNase Out, and of 1 μL Superscript III or water. Incubate 50 °C for 50 min, 70 °C for 15 min, then store at −20 °C (*see* **Notes 20** and **21**).

13. Dilute each sample to 250 μL total volume with water.

14. For each qPCR reaction use 5 μL of cDNA mixture (~8 ng) and to this add: 1 μL Taqman probe, 10 μL Taqman gene expression master mix, and 4 μL H_2O (*see* **Note 22**).

15. Analyze data to confirm expression of lentiviral vectors.

3.3.2 Flow Cytometry Analysis and Cell Sorting

1. Lift and suspend cells as described in Sect. 3.2.2 **steps 1** through 7.

2. Suspend 5×10^5 cells in 1 mL flow cytometry solution and split the cell suspension between two 1.5 mL microcentrifuge tubes and incubate at room temperature for 10 min while rocking. The remaining cells can be cryopreserved as described in Sect. 3.2.4 (*see* **Note 23**).

3. Centrifuge the cells for 4 min at $300 \times g$.

4. Aspirate the supernatant and suspend one of the cell pellets in 100 μl flow cytometry solution with diluted A2B5 conjugated antibody and the second cell pellet in 100 μl flow cytometry solution with diluted isotype control.

5. Incubate the cells at room temperature for 45 min while rocking. Keep the cells protected from light from this point onward.

6. Add 0.5 mL flow cytometry solution to each tube. Centrifuge the cells for 4 min at $300 \times g$.

7. Aspirate the supernatant carefully and suspend the cell pellets in 500 μL of OPC culture media and centrifuge for 4 min at $300 \times g$. Repeat this step once more.

8. Aspirate the supernatant and suspend the cell pellets in 200 μL of OPC culture media and strain the cells through a 5 mL round bottom polystyrene tube with 0.35 μm cell strainer snap cap.

9. Use isotype control to set gates for A2B5 positive cells

10. Sort A2B5 positive cells into 1 mL OPC Medium

11. Add 9 mL OPC Medium and spin cells at $300 \times g$ for 4 min.

12. Aspirate medium and suspend cells for further expansion as described in Sect. 3.2.2 **step 8** or aspirate medium and lyse in 1 mL TRIzol for analysis (*see* **Note 16**). See Fig. 1c and d for representative images of A2B5$^+$ iOPCs and pluripotent stem cell OPCs.

3.3.3 Differentiation to Oligodendrocytes

1. Lift and suspend cells as described in Sect. 3.2.2 **step 1** through 7.

2. Aspirate the laminin from the culture wells and immediately seed 2×10^5 cells in a Nunclon Δ-treated 6-well pretreated with poly-L-ornithine and laminin with 2 mL of OPC culture media supplemented with 200 ng/mL SHH, 100 ng/mL Noggin, 50 μm cAMP, 100 ng/mL IGF, 10 ng/mL NT-3, and 40 ng/mL T3. Or, seed 1×10^5 cells in a 12-well Nunclon Δ, poly-L-ornithine, laminin-treated with 1 mL of OPC culture

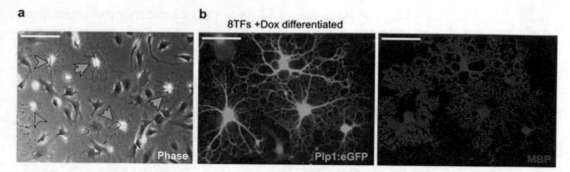

Fig. 2 (**a**) Phase-contrast image of day-21, 8TF-induced MEFs passaged into oligodendrocyte differentiation media for 3 d showing the generation of iOLs (green arrowheads), which have the distinctive multiprocessed morphology of oligodendrocytes. (**b**) Representative immunofluorescent images of iOLs differentiated from 8TF-induced MEFs, containing the Plp1-eGFP reporter and expressing the specific and defining markers of mature oligodendrocytes MBP. Scale bars, 100 μm (**a**), 25 μm (**b**)

media supplemented with 200 ng/mL SHH, 100 ng/mL Noggin, 50 μm cAMP, 100 ng/mL IGF, 10 ng/mL NT-3, and 40 ng/mL T3.

3. Incubate at 37 °C for 72 h followed by fixation with 4% PFA, (*See* Fig. 2a).

3.3.4 Immunostaining of OPC and Oligodendrocyte Markers

1. In a fume hood, aspirate cell culture media from the cells and add 4% paraformaldehyde to each well. Incubate at room temperature for 15 min while rocking.

2. In a fume hood, aspirate the paraformaldehyde and rinse the wells 3× with 1× PBS.

3. Aspirate the PBS and add 0.2% Triton X-100 solution to each well and incubate at room temperature for 10 min while rocking.

4. Aspirate the Triton X-100 solution and add blocking solution to each well. Incubate at room temperature for 1 h while rocking.

5. Aspirate and add primary antibodies diluted in blocking solution. Incubate at room temperature for 1 h while rocking or 4 °C overnight while rocking.

6. Aspirate and add 1× PBS then incubate at room temperature for 5 min while rocking. Repeat this step twice more.

7. Aspirate and add secondary antibodies diluted in blocking solution. Incubate at room temperature for 1 h while rocking. Keep the plate protected from light by wrapping the plate in aluminum foil.

8. Aspirate and add 1× PBS and incubate at room temperature for 5 min while rocking. Repeat this step twice more.

9. Aspirate and add DAPI solution diluted in 1× PBS and incubate at room temperature for 5 min while rocking.

10. Aspirate and add 1× PBS and incubate at room temperature for 5 min while rocking.

11. Aspirate and add 1× PBS. Seal plates with Parafilm, cover with aluminum foil to protect from light, and store at 4 °C. *See* Fig. 2b for an example of positive oligodendrocyte staining.

3.3.5 Global Gene Expression Analysis by RNA Sequencing

1. Isolate total RNA from A2B5 positive sorted or "Day 21" cells as described in Sect. 3.3.1 **steps 1** through **8**.

2. Prepare RNA sequencing libraries as described previously [9] (*see* **Note 24**)

3. Compare RNA sequencing results of iOPCs to those of mouse embryonic fibroblasts and bona fide OPCs. Reprogrammed iOPCs should have a distinct gene expression signature from the starting fibroblast population, similar to that of epiblast stem cell or in vivo derived OPCs (*see* **Note 25**). An example of gene expression analysis comparing pluripotent stem cell–derived OPCs, 8TF-induced MEFs (8TFs +Dox), MEFs, and uninfected MEFs can be seen in Fig. 1e.

4 Notes

1. For higher lentiviral expression efficiency it is recommended to isolate MEFs from R26-M2rtTA mice (B6.Cg-Gt(ROSA)26Sortm1(rtTA*M2)Jae/J; Jackson Laboratory) as this ensures that all MEFs used for direct reprogramming harbor the reverse tetracycline-controlled transactivator (rtTA-M2).

2. Be sure to thoroughly separate and discard neural tissue during MEF isolation to avoid contamination before reprogramming. It is highly recommended to stain each batch of MEFs for markers of oligodendrocyte progenitor cells (Olig2, Sox10, and Nkx2.2) to ensure no contamination of in vivo derived OPCs before reprogramming. *See* Fig. 3 for schematic representation of MEF isolation and control staining.

3. Nunclon Δ-treated individual 6-wells are 9.6 cm² while Nunclon Δ-treated individual 12-wells are 3.5 cm².

4. An HEK 293T cell line should be used, not a parental HEK 293 cell line, to ensure sufficient viral production.

5. Lentiviral vectors have been deposited with Addgene and can be ordered using the following ID numbers: *Nkx6.2* (64846), *Olig2 (64842)*, *Sox10 (46536)*, *Olig1* (64841), *Nkx2.2* (64840), *Myt1* (64845), *St18* (64844), and *Myrf* (64843).

6. If not directly reprogramming MEFs isolated from R26-M2rtTA mice, lentivirus will need to be produced using a pLVX-Tet-On viral plasmid (Clontech). Direct reprogramming efficiency will be lower if using pLVX-Tet-On lentiviral

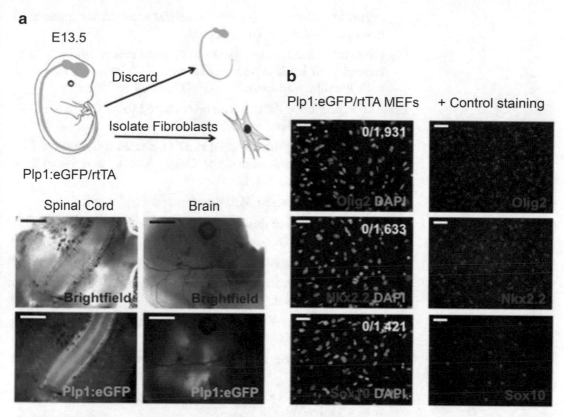

Fig. 3 (**a**). Diagram of the strategy used to isolate MEFs devoid of neural tissue from E13.5 Plp1:eGFP/rtTA mice along with brightfield and fluorescent images of the embryos. (**b**) Isolated Plp1:eGFP/rtTA, which are used as the starting cell source for reprogramming experiments, should be negative oligodendrocyte lineage markers. Positive control pluripotent stem cell derived OPCs were stained with MEFs to ensure function of each antibody (Olig2, Nkx2.2, and Sox10). Scale bars, 1 mm (**a**), 50 μm (**b**)

supernatant as opposed to starting MEFs harboring the reverse tetracycline-controlled transactivator (rtTA-M2).

7. This probe spans an exon junction and will therefore detect endogenous Olig2 expression.

8. At this point the virus is active and should be handled with care. All equipment and plastic ware that comes into direct contact with the lentivirus supernatant should be rinsed with 100% bleach solution.

9. It is not recommended to seed MEFs for lentiviral infection directly from frozen stocks, but instead to thaw and culture for at least one passage. This ensures an accurate number of live cells are seeded to each well.

10. If infecting MEFs with only *Sox10*, *Olig2*, and *Nkx6.2* add equal volumes (333 μL) of fresh lentiviral supernatant to each 6-well. In addition, if the MEFs do not harbor the reverse tetracycline-controlled transactivator (rtTA-M2) adjust the viral

infection volume to include lentiviral supernatant generated from pLVX-Tet-On viral plasmid.

11. Once fully dried, poly-L-ornithine coated plates can be further treated with laminin and seeded with iOPCs or can be sealed with Parafilm and stored at −20 °C for up to 1 month.

12. At this point 5×10^4 cells can be collected for quantitative PCR analysis as described in Sect. 3.3.1.

13. After incubating the 12-wells at 37 °C for 24 h they should be fixed and immunostained for Olig2, Sox10, and Nkx6.2 as described in Sect. 3.3.4.

14. At this point infected MEFs are considered "day 3."

15. Working in a space designated for RNA is recommended. This reduces the risk of contamination.

16. Cell lysate can be stored in TRIzol at 80 °C for several months.

17. If isolating RNA from samples that have been stored at 80 °C allow the cell lysate to thaw on ice.

18. Centrifuge Phase-Lock tubes at room temperature prior to use to pellet the wax.

19. This step is the only step in which the centrifugation is carried out at 4 °C.

20. When isolating RNA from multiple samples it is best to prepare a Master Mix.

21. It is important to prepare a negative Reverse Transcription control where Superscript III is replaced with water in the Master Mix.

22. When setting up the qPCR reactions it is best to have at least three technical replicates for downstream data analysis.

23. Use approximately 1×10^5 cells for isotype control and the remaining cells for analysis or sorting.

24. Perform RNA Sequencing library construction as in Factor et al. Sect. 3.3.2 **steps 15** through **20** [9].

25. RNA seq data sets: mouse epiblast stem cell derived OPCs (GSM1557625), mouse postnatal day 7 brain OPCs(GSM1557624), Mouse Embryonic Fibroblast (GSM929719) [10, 11].

References

1. Franklin RJ, Ffrench-Constant C (2008) Remyelination in the CNS: from biology to therapy. Nat Rev Neurosci 9(11):839–855. doi:10.1038/nrn2480

2. Windrem MS, Schanz SJ, Guo M, Tian GF, Washco V, Stanwood N, Rasband M, Roy NS, Nedergaard M, Havton LA, Wang S, Goldman SA (2008) Neonatal chimerization with human glial progenitor cells can both remyelinate and rescue the otherwise lethally hypomyelinated shiverer mouse. Cell Stem Cell 2(6):553–565. doi:10.1016/j.stem.2008.03.020

3. Goldman SA, Nedergaard M, Windrem MS (2012) Glial progenitor cell-based treatment

and modeling of neurological disease. Science 338(6106):491–495. doi:10.1126/science.1218071

4. Wang S, Bates J, Li X, Schanz S, Chandler-Militello D, Levine C, Maherali N, Studer L, Hochedlinger K, Windrem M, Goldman SA (2013) Human iPSC-derived oligodendrocyte progenitor cells can myelinate and rescue a mouse model of congenital hypomyelination. Cell Stem Cell 12(2):252–264. doi:10.1016/j.stem.2012.12.002

5. Sim FJ, McClain CR, Schanz SJ, Protack TL, Windrem MS, Goldman SA (2011) CD140a identifies a population of highly myelinogenic, migration-competent and efficiently engrafting human oligodendrocyte progenitor cells. Nat Biotechnol 29(10):934–941. doi:10.1038/nbt.1972

6. Najm FJ, Zaremba A, Caprariello AV, Nayak S, Freundt EC, Scacheri PC, Miller RH, Tesar PJ (2011) Rapid and robust generation of functional oligodendrocyte progenitor cells from epiblast stem cells. Nat Methods 8(11):957–962. doi:10.1038/nmeth.1712

7. Najm FJ, Lager AM, Zaremba A, Wyatt K, Caprariello AV, Factor DC, Karl RT, Maeda T, Miller RH, Tesar PJ (2013) Transcription factor-mediated reprogramming of fibroblasts to expandable, myelinogenic oligodendrocyte progenitor cells. Nat Biotechnol 31(5):426–433. doi:10.1038/nbt.2561

8. Chenoweth JG, Tesar PJ (2010) Isolation and maintenance of mouse epiblast stem cells. Methods Mol Biol 636:25–44. doi:10.1007/978-1-60761-691-7_2

9. Factor DC, Najm FJ, Tesar PJ (2013) Generation and characterization of epiblast stem cells from blastocyst-stage mouse embryos. Methods Mol Biol 1074:1–13. doi:10.1007/978-1-62703-628-3_1

10. Najm FJ, Madhavan M, Zaremba A, Shick E, Karl RT, Factor DC, Miller TE, Nevin ZS, Kantor C, Sargent A, Quick KL, Schlatzer DM, Tang H, Papoian R, Brimacombe KR, Shen M, Boxer MB, Jadhav A, Robinson AP, Podojil JR, Miller SD, Miller RH, Tesar PJ (2015) Drug-based modulation of endogenous stem cells promotes functional remyelination in vivo. Nature. doi:10.1038/nature14335

11. Yue F, Cheng Y, Breschi A, Vierstra J, Wu W, Ryba T, Sandstrom R, Ma Z, Davis C, Pope BD, Shen Y, Pervouchine DD, Djebali S, Thurman RE, Kaul R, Rynes E, Kirilusha A, Marinov GK, Williams BA, Trout D, Amrhein H, Fisher-Aylor K, Antoshechkin I, DeSalvo G, See LH, Fastuca M, Drenkow J, Zaleski C, Dobin A, Prieto P, Lagarde J, Bussotti G, Tanzer A, Denas O, Li K, Bender MA, Zhang M, Byron R, Groudine MT, McCleary D, Pham L, Ye Z, Kuan S, Edsall L, Wu YC, Rasmussen MD, Bansal MS, Kellis M, Keller CA, Morrissey CS, Mishra T, Jain D, Dogan N, Harris RS, Cayting P, Kawli T, Boyle AP, Euskirchen G, Kundaje A, Lin S, Lin Y, Jansen C, Malladi VS, Cline MS, Erickson DT, Kirkup VM, Learned K, Sloan CA, Rosenbloom KR, Lacerda de Sousa B, Beal K, Pignatelli M, Flicek P, Lian J, Kahveci T, Lee D, Kent WJ, Ramalho Santos M, Herrero J, Notredame C, Johnson A, Vong S, Lee K, Bates D, Neri F, Diegel M, Canfield T, Sabo PJ, Wilken MS, Reh TA, Giste E, Shafer A, Kutyavin T, Haugen E, Dunn D, Reynolds AP, Neph S, Humbert R, Hansen RS, De Bruijn M, Selleri L, Rudensky A, Josefowicz S, Samstein R, Eichler EE, Orkin SH, Levasseur D, Papayannopoulou T, Chang KH, Skoultchi A, Gosh S, Disteche C, Treuting P, Wang Y, Weiss MJ, Blobel GA, Cao X, Zhong S, Wang T, Good PJ, Lowdon RF, Adams LB, Zhou XQ, Pazin MJ, Feingold EA, Wold B, Taylor J, Mortazavi A, Weissman SM, Stamatoyannopoulos JA, Snyder MP, Guigo R, Gingeras TR, Gilbert DM, Hardison RC, Beer MA, Ren B (2014) A comparative encyclopedia of DNA elements in the mouse genome. Nature 515(7527):355–364. doi:10.1038/nature13992

Isolation and Culture of Oligodendrocyte Precursor Cells from Prenatal and Postnatal Rodent Brain

Danyang He, Bradley Meyer, and Q. Richard Lu

Abstract

Oligodendrocytes, the myelinating cells in the vertebrate central nervous system, electrically insulate axons and facilitate rapid propagation of action potentials. The isolation and transplantation of oligodendroglial precursor cells (OPCs) could provide a powerful means to characterize their differentiation properties and potential and enhance myelin repair. Preparation and maintenance of rodent OPCs have been a challenge due to difficulties in obtaining a sufficient quantity. Here, we describe step-by-step and efficient methods to prepare highly enriched OPCs from rat and mice. Isolated OPCs can be induced to differentiate into oligodendrocytes and express most, if not all, of oligodendrocyte-associated genes in vivo. The ability to isolate and culture OPCs should facilitate studies on oligodendrocyte lineage progression and their utility in myelin repair after injury.

Key words Oligodendrocyte lineage, OPC, Mouse cortex, Rat cortex, Mixed glial culture, Mechanical dissociation, Immunopanning, Oligodendrosphere, Neurosphere

1 Introduction

Oligodendrocytes, the myelinating glia of the central nervous system (CNS), play a crucial role in facilitating the rapid conduction of action potentials [1], supporting axonal survival [2, 3], and motor and cognitive functions [4, 5]. Oligodendrocytes are generated from their precursor cells (OPCs), which proliferate and migrate throughout the central nervous system during late embryonic development, and later differentiate into mature, myelinating oligodendrocytes [6]. Several distinct stages have been identified for oligodendroglial lineage cells during oligodendrocyte maturation in vitro [7, 8]. These include proliferating OPCs, characterized by expression of progenitor cell markers such as the platelet-derived growth factor alpha receptor (PDGFRα) or NG2 or O4 with a bipolar or tripolar morphology [9], and mature oligodendrocytes, expressing myelin-specific proteins such as myelin basic protein. Upon transplantation into the CNS of hypomyelinated hosts like

Amit K. Srivastava et al. (eds.), *Stem Cell Technologies in Neuroscience*, Neuromethods, vol. 126, DOI 10.1007/978-1-4939-7024-7_6, © Springer Science+Business Media LLC 2017

Shiverer mice, oligodendrocyte precursors can migrate a significant distance, and give rise to a large number of myelinating oligodendrocytes [10], which ensheath axons [11]. Development of simple methods for the isolation and purification of workable quantities of OPCs will not only aid in efforts to better understand oligodendrocyte differentiation, function and axon–oligodendroglia interactions but also provide an indispensable tool for myelin repair. The current methods for isolation of OPCs from the immature brain are through immunopanning [12–14], fluorescence-activated cell sorting (FACS) [8], exploiting cell surface-specific antigens, differential centrifugation [11, 15], or shaking methods based on differential adherent properties of glia [16–18], which permit the separation of rat OPCs from the astroglial cells in the mixed glial culture by shearing forces [16]. The fact that multipotent neural stem/progenitor cells can give rise to oligodendroglial lineage restricted progenitors suggested a new avenue for the generation of pure OPCs from neural stem/progenitor cells. Several methods have been described to generate self-renewing OPCs from neural stem cells in different species such as dog and rat [19–21].

In this protocol, we describe simple and efficient methods to isolate a large, pure population of OPCs from rats and mice. The first protocol allows isolation of rat OPCs using a selective detachment procedure [16] with modifications [22]; and the second protocol is for the generation of OPCs from oligodendrospheres from prenatal animals [18] or by immunopanning from neonatal mice. The OPCs isolated by the procedures described below can be induced to differentiate into immature oligodendrocytes and then into mature oligodendrocytes. These methods will facilitate the in vitro use of OPCs to study, for example, effects of various molecules on OPC differentiation and axon–oligodendroglia interactions.

2   Materials and Methods

2.1   Reagents

Dulbecco's Modified Eagle Media (DMEM, Invitrogen/Gibco 11960) without L-glutamine and sodium pyruvate.

DMEM/F12 (Gibco 11330-032)

Fetal bovine serum (Hyclone SH300700)

L-glutamine (Sigma G8540)

Sodium pyruvate (Sigma P2256)

Bovine serum albumin (Sigma A9647)

Tris base (Fisher Scientific BP152)

Apo-transferrin (Sigma T2252)

Insulin (Sigma I6634)

Sodium selenite (Sigma S5261)

D-biotin (Sigma B4501)

Hydrocortisone (Sigma H0888)

Human PDGF-AA (Peprotech 100-13A)

Basic FGF (Peprotech 100-18B)

Human recombinant epidermal growth factor (EGF, Peprotech, 100-15)

Forskolin (Sigma F6886)

Recombinant Human Neurotrophin-3 (NT3, Peprotech 450-03)

Recombinant Human CNTF (Peprotech 450-13)

Poly-DL-ornithine (Sigma P0421)

Poly-D-lysine (Sigma P0899)

Penicillin/streptomycin (Invitrogen 15140)

Trace element B (Cellgro 99-175-CI)

Hanks Balanced Salt Solution (HBSS, Invitrogen 14025)

DNase (Sigma D5025)

Trypsin (Sigma T1426)

Trypan blue (Sigma T8154)

Phenylmethylsulfonyl fluoride (PMSF, Sigma P7626-5G)

Trypsin Inhibitor, Ovomucoid Source (Ovo, Worthington LS003086)

Papain (Worthington LS003118)

L-cysteine hydrochloride (Sigma C1276)

Earle's Balanced Salt Solution (EBSS, Invitrogen 14155063)

D-PBS with calcium and magnesium (Invitrogen 14040)

2.2 Equipment

Humidified tissue incubator (37 °C, 5% CO_2)

Biosafety cabinet

Dissecting microscope

Water bath at 37 °C

Microdissecting instruments (sterilized):

– Small dissecting scissors

– Dumont forceps, straight and curved

– Curved microdissecting scissors

– Spatula

Tabletop centrifuge

Orbital shaker at 37 °C (Barnstead digital orbital shaker, Cat. SHKE2000)

37 °C oven for orbital shaker (Bellco benchtop incubator)

15 mL plastic conical tubes (Falcon 352097)

50 mL plastic conical tubes (Falcon 352070)

75 cm² tissue culture flask

96 well, 24-well tissue culture plates

Sterile medium filters

10 cm petri dish (Fisher 08-757-13)

70 μm cell strainer (Falcon 352350)

15–20 μm sterile screening fabric nylon mesh (Sefar America cat. No. 03-20/14)

2.3 Media and Solutions

DMEM20S: DMEM, 4 mM L-glutamine, 1 mM pyruvate, 20% fetal bovine serum, 50 U/mL penicillin, and 50 μg/mL streptomycin

Basal chemically defined medium (BDM): DMEM, 4 mM L-glutamine, 1 mM pyruvate, 0.1% BSA, 50 μg/mL apo-transferrin, 5 μg/mL insulin, 30 nM sodium selenite, 10 nM D-biotin, 10 nM hydrocortisone.

Rat OPC medium: BDM containing 10 ng/mL PDGF-AA and 10 ng/mL bFGF.

Critical step: Growth factors are prepared as 1000× stocks and stored in aliquots at −80 °C. They are added to BDM just prior to OPC plating or medium change.

Neurosphere medium: DMEM/F12 supplemented with insulin (25 μg/mL), transferrin (100 μg/mL), progesterone (20 nM), putrescine (60 μM), sodium selenite (30 nM). *Neurosphere growth medium:* Neurosphere medium plus 20 ng/mL bFGF and 20 ng/mL EGF.

B104 growth medium: (for the growth of B104 neuroblastoma cells) DMEM/F12 supplemented with 10% FBS.

B104-N2 medium: DMEM/F12 supplemented with 1× N_2.

Leibovitz's L-15 medium 1× (Invitrogen 21083) 1× Puck's BSS: a calcium- and magnesium-free balanced salt solution.

10× Puck's BSS: NaCl 80 g, KCL 4 g, $Na_2HPO_4.7H_2O$ 0.9 g, KH_2PO_4 0.4 g, and glucose 10 g in 1000 mL triple distilled water and filter-sterilized.

Working solution, 1× Puck's BSS: dilute stock solution 1:10 with sterile triple-distilled water.

Preparation of B104 neuroblastoma conditioned medium (B104 CM):

- Culture B104 neuroblastoma cells in neuroblast growing medium until confluent.

- Wash with 1× Puck's BSS, and feed with B104-N_2 medium.

- After 4 days, collect the medium, add PMSF to a final concentration 1 μg/mL, mix quickly.

- Centrifuge at 2000 × g for 30 min.

- Filter the supernatant with a 0.22 μm pore-size filter system, and obtain B104 CM.

  Note: The B104 CM can be collected and stored in −80 °C for later use. To minimize culture to culture variation we generally use the CM from the same batch. Freeze–thaw cycles should be minimal.

  Antibodies:
  Anti-GalC antibody (Millipore MAB344).
  Anti-O4 antibody (Millipore MAB345).
  AffiniPure F(ab')$_2$ Fragment Goat Anti-Mouse IgG + IgM (H+L) (Jackson ImmunoResearch 115-006-068).
  Anti-RAN-2 (ATCC® TIB-119™).
  Papain buffer: 1× EBSS solution with 1 mM MgSO$_4$, 2 mM EGTA, and 0.36% glucose.
  Mouse OPC medium: BDM containing 5 μM Forskolin, 10 ng/mL CNTF, 10 ng/mL PDGF-AA, 10 mg/mL NT3 and 2× B27.
  Oligodendrosphere medium: 7 parts of neurosphere medium: 3 parts of B104CM (7:3 = vol:vol).
  DNase: 20× stock 200 μg/mL in HBSS, store in aliquots at −20 °C.
  Trypsin: 25× stock 0.25% in HBSS, store in aliquots at −20 °C.
  Lo Ovo (10× stock): 100 mg BSA and 100 mg Ovo dissolved in 20 mL D-PBS, filter-sterilize and store in aliquots at −20 °C.
  High Ovo (6× stock): 1200 mg BSA and 1200 mg Ovo dissolved in 20 mL D-PBS (dissolve at 4 °C overnight if necessary), filter-sterilize and store in aliquots at −20 °C.
  5% BSA solution (25× stock): 2.5 g BSA to 50 mL D-PBS, filter-sterilize and store in aliquots at −20 °C.
  Panning buffer: dilute 800 μL 5% BSA stock in 20 mL D-PBS. Prepare this solution just prior to dissection.
  Tris–HCl solution (50 mM): 6 g Tris base in 1000 mL triple distilled water, adjust pH to 9.5 and filter-sterilize.
  Animals used: C57B6/J mice or Sprague-Dawley rats.

2.4 Isolation and Purification of OPCs from Neonatal Rats

2.4.1 Procedures

Coating flasks with poly-D-lysine

1. Dilute 100× stock of poly-D-lysine (10 mg/mL in 0.5% BSA in PBS, stored as aliquots at −20 °C) with PBS and filter to sterilize.

2. Coat culture flasks and plates with the 1× poly-D-lysine for 1–2 h in 37 °C incubator.

3. Remove coating solution and wash twice with sterile ddH$_2$O.

4. Store with caps screwed tight at RT.

Coating plates with poly-D, L-ornithine

1. Dilute 100× stock of poly-D, L-ornithine (5 mg/mL in PBS, stored in aliquots at −80 °C) with PBS and filter to sterilize.

2. Add coating solution to culture plates and incubate for 1–2 h at 37 °C.

3. Remove solution and wash twice with ddH$_2$O.

4. Store at RT.

   Tip: coated flask and plates can be stored up to 4 weeks at 4 °C.

Dissection and mixed glial culture

1. In the biosafety cabinet, pour ice-cold PBS into two 10 cm petri dishes placed in tray filled with ice.

2. Decapitate P1-2 rat pup with a large scissors (1 L approx. 10 pups at a time) and place in cold DPBS.

3. Proceed and place 1 head at a time in a petri dish containing clean 70% ethanol briefly and then in petri dish containing cold DPBS. Then, transfer it into another petri dish containing ice-cold PBS. Repeat for the remaining heads in the litter (approximately 10–15 pups can be processed at a time). Change to a clean petri dish containing cold PBS if necessary.

4. One head at a time in a clean petri dish, use curved microdissecting scissors to gently cut the skin along the midline, and then cut the skull.

   Note: The tip of the scissors should point away from brain to avoid damaging the brain.

5. Make two lateral cuts at the base of the skull by inserting the scissors where the spinal cord was severed at the foramen magnum.

6. Fold back the two sides of the skull with forceps, and scoop out the brain, and subsequently cut off the cerebellum with a spatula. Place the forebrain in a clean petri dish containing PBS on ice.

7. Repeat procedure for remaining heads.

8. Remove the meninges from the forebrain with forceps under dissection microscope.

9. Place all meninges-free brains into one clean petri dish on ice.

10. Transfer the meninges-free cortices from three brains into one 1.5 mL tube and add 1 mL cold DMFM/F12 with 15% FBS.

11. Triturate cortices gently by passing the tissues sequentially through 18 (twice), 23 (twice), and 25 (once) gauge needles with a 1 mL syringe.

12. Repeat to triturate all cortices, and combine the cells suspensions.

13. Spin down ≈800 RPM.

14. Carefully aspirate supernatant with Pasteur pipette.

15. Add 20 mL DMEM20S to the tube.

16. Triturate tissue with glass 10 mL pipette until nearly homogenous.

17. Pass suspension (avoid pellet) through a 70 μm Nylon cell strainer and collect in a 50-mL tube.

18. Add 20 mL DMEM20S to settled tissue and triturate again.

19. Combine filtered suspension in a T75 poly-D-lysine coated flask and volume to 100 mL with DMEM20S.

20. Plate 12 mL cell suspension per poly-D-lysine coated flask.

21. Spread media over entire flask surface and date flask.

22. Feed every 2–3 days with complete medium changes with 10 mL DMEM20S.

23. At ~10 days after plating, mixed glial cultures will be confluent and can be shaken to obtain OPC cultures.

OPC Isolation and Plating

1. Remove culture flasks from incubator.

2. Screw caps on tight.
   Note: Viability of OPCs is unaffected by the closed environment during the shaking procedure.

3. Secure flasks to a Styrofoam board on a horizontal orbital shaker (autoclave tape holds better than regular paper tape).

4. Pre-shake the flasks on the shaker for 1 h at 180 rpm and 37 °C to remove microglial cells.

5. Discard medium from flasks by aspiration.

6. Add 12 mL DMEM20S to each flask and incubate cells at 37 °C for 3–4 h to allow the CO_2 to re-equilibrate.

7. Secure flasks to orbital shaker at 37 °C and 200 rpm overnight (≈18–22 h.)

8. Remove flasks from shaker and sterilize the surface with 70% ethanol.

9. Collect cell suspension from each flask, pass through 70 μm nylon cell strainer over a clean 50 mL conical tube. Transfer cells to untreated petri dish (1 flask per petri dish).
   Note: Only petri dishes not treated for cell culture should be used or else OPCs will tend to attach to dishes, resulting in very low yield of OPCs.

10. Add 10 mL DMEM20S to flasks needing a second shake 1 week later, and discard flasks that have been shaken twice.
    Note: Mixed glia can be shaken for the second time if one continues to culture them for an additional week after the first shake.

11. Incubate petri dishes for 30–60 min in incubator for differential adhesion of contaminating microglia and astrocytes.

12. Gently swirl the petri dish and remove all medium into a 50-mL tube.

Isolation of rat OPCs

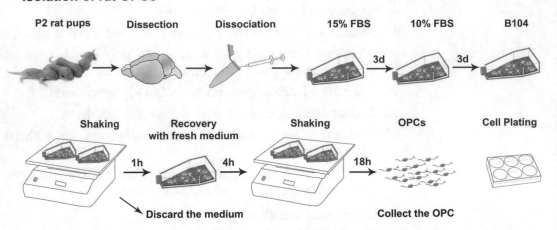

Fig. 1 Isolation of oligodendrocyte precursor cells from neonatal rats by shaking methods

13. Flame a clamp and attach the screening pouch (15–20 µm pore size) to the clamp.

14. Pass the cell suspension slowly through the sieves over a clean 50 mL tube.

15. Centrifuge cell suspension for 10 min at 800 RPM.

16. Carefully remove supernatant by pipette without dislodging the pellet. Suspend pellet in small amount of remaining medium.

17. Count living cells with trypan blue exclusion assay.

18. Dilute OPC suspension with the OPC medium to desired concentration and plate into poly-D, L-ornithine-coated plates at density of $1 \times 10^4/cm^2$.

19. Feed cells every other day with half change of medium with BDM containing 2× of PDGF and FGF. Cells are primarily A2B5$^+$, O4$^+$, O1$^-$, and MBP$^-$ oligodendrocyte precursors after 7 days in vitro (Fig. 1). Contaminating of microglia and astrocytes are routinely less than 2–3% each.

2.5 Isolation and Purification of OPCs from Neonatal Mice

2.5.1 Procedures

Day1:
Coating plates or coverslips with poly-D, L-ornithine

1. Dilute 100× stock of poly-D, L-ornithine (5 mg/mL in PBS, stored in aliquots at −80 °C) with PBS and filter to sterilize.

2. Add coating solution to culture plates and incubate for 1–2 h at 37 °C.

3. Remove solution and wash twice with ddH$_2$O.

4. Store at RT.

Tip: coated flask and plates can be stored up to 4 weeks at 4 °C.

Coating three panning dishes per prep with 2nd antibody

1. Make 30 mL of 10 μg/mL goat anti-mouse IgG+IgM solution in sterile 50 mM Tris–HCl (pH = 9.5). Coat each 100 mm petri dish with 10 mL antibody-Tris solution.

2. Incubate panning plates at 4 °C overnight.
   Note: Antibody stock solution is subject to change and check the vial/tube of stock before use.

Day2:
Before dissection:

1. Pre-warm papain buffer in 37 °C water bath.

2. Defrost and make up primary antibodies Ran2, GalC, and O4 (see steps below).

3. Set up panning dishes.

   • Wash goat anti-mouse IgG+IgM coated panning dishes three times.

   • Add primary antibodies onto panning dishes:

   Ran2: 7 mL antibody + 3 mL panning buffer
   GalC: 5 mL antibody + 5 mL panning buffer
   O4: 5 mL antibody + 5 mL panning buffer

   • Let the dishes sit at RT until used.

Dissection

1. Disinfect the dissection area with 70% ethanol.

2. Cut off heads of mouse pups (P5-P6). Remove skin and skull.

3. Remove cortical hemispheres and place in 100 mm dish with 20 mL PBS.

4. Remove pial membranes, transfer cortical hemispheres to 60 mm dishes with ~3–5 mL PBS (~6 hemisphere per prep).

5. Remove PBS from 60 mm dish without removing any tissue and dice brains into ~1–2 mm chunks with #10 scalpel blade.

Dissociation

1. Prepare Papain solution by adding 200 units papain and 2 mg L-cysteine to 10 mL pre-warmed papain buffer, vortex and filter to sterilize.
   Note: Papain concentration will change per batch so check each vial for stock concentration.

2. Add 2500 U (200 μL per prep) DNase to filtered papain solution.

3. Transfer minced brains to 10 mL papain solution in 50 mL conical tube and incubate at 37 °C for 30 min with shaking at 15 min intervals.

Trituration

1. Prepare 10 mL Lo Ovo by diluting 10× stock in D-PBS and adding 2500 U DNase. Prepare 6 mL Hi Ovo by adding 6× stock in DBPS.
   Tip: Make Lo Ovo solution during dissociation.

2. Remove as much papain solution as possible. Add 2 mL Lo Ovo, allow tissue chunks to settle.

3. Gently remove 2 mL supernatant. Add another 2 mL Lo Ovo and gently triturate tissue chunks 5–6 times with a 5-mL pipette.

4. Allow cells to settle, remove 1 mL from the top of the suspension and place in a 15-mL tube.

5. Add 1 mL Lo Ovo, triturate with a 5-mL pipette, let settle, save 1 mL supernatant from the top and add to the previous 1 mL supernatant.

6. Add 1 mL Lo Ovo and triturate with a P-1000 pipetman.

7. Replace the previous 2 mL of saved suspension to the original tube and spin cells at 1300 RPM for 15 min.

8. Remove the supernatant and suspend the cell pellet in 2 mL Hi Ovo.

9. Add remaining Hi Ovo and spin at 1300 RPM for 15 min.

10. Remove the supernatant and resuspend the pellet in 8 mL of panning buffer.

11. Pre-wet a nylon mesh filter (40 μm) with 2 mL panning buffer and filter the cell suspension into a 50-mL conical tube. With the remaining 2 mL of panning buffer, rinse the filter.

Panning

1. Immediately prior to use of each panning dish use, wash the dish three times with D-PBS.

2. Apply cell suspension to the panning plates for the indicated incubation time with gentle shaking every 15 min in the following order:
   - Ran2 plate, 30 min
   - GalC plate, 30 min
   - O4 plate, 45 min

Rinse plates with 3 mL D-PBS in between each transfer.

Isolation of mouse OPCs

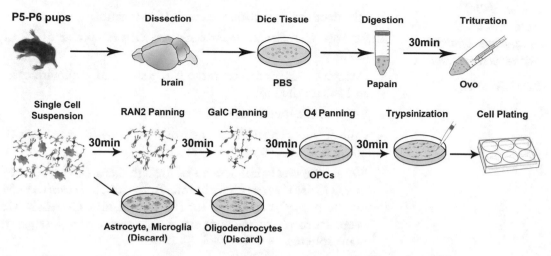

Fig. 2 Isolation of oligodendrocyte precursor cells from neonatal mice by immunopanning methods

Trypsinization, collection and plating of OPCs

1. Take the supernatant off the O4 plate and wash the plate three times with D-PBS and wash three times with DPBS. Make sure all non-adherent cells are removed.

2. Add 5 mL pre-warmed 0.25% trypsin and place in a 37 °C incubator for no longer than 5 min (~30 s). Check for the detachment of cells by gently tapping the flask.

3. Add 5 mL 30% FBS in D-PBS to inactivate the trypsin. Dislodge OPCs from the plate surface by tipping the dish and squirting the exposed surface with a P-1000 pipetman.

4. Transfer the suspension to a 50-mL tube. Add another 5 mL 30% FBS in D-PBS, check plate for adhered cells and wash off remaining attached cells with P-1000 pipet. Combine cell suspensions in the 50-mL tube.

5. Spin the cells at 1,300 RPM (390 × g) for 20 min.

6. Resuspend cells in DMEM/F12 with 10% FBS.

7. Count cells, dilute the OPCs to desired density and seed cells to poly-D, L-ornithine-coated coverslips or wells.

8. Incubate the cells in a tissue culture incubator at 37 °C for 30–60 min. Then change to mouse OPC medium.

9. Feed cells every other day with half change of medium for mouse OPC expansion (Fig. 2).

**2.6   Isolation
of Mouse OPCs
from Neural
Progenitors-Derived
Oligodendrospheres**

2.6.1   Procedures

Dissection of embryonic cortices

1. Disinfect the dissection area with 95% ethanol.

2. Remove the mouse or rat embryos for example at e14.5 or e17.5.

3. Decapitate the embryos, remove the skin and skull with care, and isolate the brain.

4. Remove the meninges with fine forceps.

5. Collect the cortex in 2 mL neurosphere growth medium with 20 ng/mL EGF and 20 ng/mL bFGF.

   Note: The biological activity of growth factors in serum free media (SFM) decreases with time. Add growth factors to SFM on the day of preparation for the best results. Generally, we keep neurosphere medium up to 7 days for the passage of neurospheres.

6. Dissociate the cortex by mechanical trituration with a fire polished glass Pasteur pipette (about 35 strokes) until a turbid suspension with a few pieces is achieved.
   Set it on ice for 2 min.

7. Remove the supernatant and filter through a 50 μm nylon mesh to isolate single cells. Add another 2 mL of neurosphere culture medium and mix well.

8. Count the cells and add 5×10^4 cells/mL cell suspension to each well of a six-well plate.

Generation of Oligodendrospheres from Neurospheres

1. Culture the cells (5×10^4/mL) in the uncoated plate in the neurosphere medium with 20 ng/mL bFGF and 20 ng/mL EGF.

2. Replace neurosphere medium every other day.

3. After the formation of the neurospheres (Usually 3–5 days for the embryonic tissue and for the neonatal it requires around 10 days), the mixed culture is ready for oligodendrocyte culture.

4. Gradually change the EGF/bFGF-containing neurosphere medium to B104 CM– containing oligodendrosphere medium by replacing one-fourth of the former medium with the latter medium every other day.

   Note: During the transition period (1–2 weeks), the number and size of spheres remains the same. However, the GFP fluorescence does increase significantly.

5. After 2 weeks, the size and the number of spheres begins to increase at a greater rate.

   Note: Almost 90% of spheres show high GFP fluorescence as detected from PDGFαR-GFP+ OPC reporter. Since the cells are PDGFαR-GFP+, signifying oligodendrocyte precursors, the spheres are now referred as oligodendrospheres.

Culturing of Oligodendrospheres

1. (a) Mechanically dissociate oligodendrospheres, and filter through a 50 μm nylon mesh to obtain single cells.
   (b) Alternatively, dissociate by using enzymatic treatment with trypsin.

   Incubate oligodendrospheres with 0.05% trypsin at 37 °C for 5 min.

   Neutralize with culture medium such as oligodendrosphere medium.

   Spin down the cells.

   Pass through a 50 μm nylon mesh to obtain single cells.

2. Culture the cells in oligodendrosphere medium on uncoated plate. Oligodendrospheres form again after 5–12 days.

   TIP: To generate bulk density of neurospheres, plate the cells in higher density (e.g., 100 cells/μL). Higher density plating will yield more neurospheres in a short time.

   Caution:

   While passaging oligodendrospheres some spheres tend to attach to the bottom of the plate and assume typical early oligodendrocyte morphology. Generally, only the free-floating spheres are taken for passaging. Avoid excessive formation of air bubbles while doing mechanical dissociation of neurospheres or oligodendrospheres as it will reduce cell viability.

3 Troubleshooting

Culturing of neurospheres

A low yield of neurospheres is often a result of culturing of very high cell density, the dead and dying cells will prevent the generation of neurospheres.

Oligodendrosphere generation: occasionally, few oligodendrospheres without detectable GFP expression were observed. These spheres can be removed by a pipette to keep nearly 100% pure PDGFαR-GFP positive oligodendrospheres as described previously [18].

4 Discussion

These protocols have been tested and adjusted for isolating OPCs from neonatal rats and mice. Isolated OPCs can be expanded in serum-free Sato media containing mitogens PDGFAA and FGF or differentiated into mature oligodendrocytes in the presence of T3, and/or withdrawing mitogens). Serum-free Sato media are required to prevent OPCs differentiate into type-2 astrocytes [23]. OPC differentiation in culture recapitulates to a large extent the differentiation of their in vivo counterparts. Thus, the OPC culture can be a suitable system to dissect the molecular mechanisms of OL differentiation and maturation as described by others [24–27]. In addition, isolated OPCs can also be used to study axonal-glial interaction in coculture system by plating together with retinal ganglion axons [28, 29]. A large quantity of OPCs can be isolated from a smaller number of rat pups compared with mouse pups. Rat OPCs are easier for expansion and manipulations such as transfection and pharmacological treatment. In contrast, mouse OPCs are relatively difficult to expand and they have tendency to differentiate. Given many of studies using transgenic, knockout, or conditional knockout animals, isolation of mouse OPCs would facilitate the functional study of the genes of interest and genomic profiling studies [30].

Acknowledgment

Authors would like to thank Dr. Jianrong Li and Dr. Jonah Chan for sharing the rat and mouse OPC isolation protocols, respectively. This study was funded in part by grants from the US National Institutes of Health R01NS072427 and R01NS075243 to QRL and the National Multiple Sclerosis Society (NMSS-4727) to QRL.

References

1. Miller G (2005) Neuroscience. The dark side of glia. Science 308(5723):778–781
2. Mar S, Noetzel M (2010) Axonal damage in leukodystrophies. Pediatr Neurol 42(4):239–242. doi:10.1016/j.pediatrneurol.2009.08.011. S0887-8994(09)00399-3 [pii]
3. Trapp BD, Peterson J, Ransohoff RM, Rudick R, Mork S, Bo L (1998) Axonal transection in the lesions of multiple sclerosis. N Engl J Med 338(5):278–285
4. Molofsky AV, Kelley KW, Tsai HH, Redmond SA, Chang SM, Madireddy L, Chan JR, Baranzini SE, Ullian EM, Rowitch DH (2014) Astrocyte-encoded positional cues maintain sensorimotor circuit integrity. Nature 509(7499):189–194. doi:10.1038/nature13161. nature13161 [pii]
5. McKenzie IA, Ohayon D, Li H, de Faria JP, Emery B, Tohyama K, Richardson WD (2014) Motor skill learning requires active central myelination. Science 346(6207):318–322. doi:10.1126/science.1254960
6. Baumann N, Pham-Dinh D (2001) Biology of oligodendrocyte and myelin in the mammalian central nervous system. Physiol Rev 81(2): 871–927
7. Pfeiffer SE, Warrington AE, Bansal R (1993) The oligodendrocyte and its many cellular processes. Trends Cell Biol 3(6):191–197
8. Gard AL, Williams WC 2nd, Burrell MR (1995) Oligodendroblasts distinguished from O-2A glial progenitors by surface phenotype (O4+GalC-) and response to cytokines using signal transducer LIFR beta. Dev Biol 167(2):596–608
9. Nishiyama A, Suzuki R, Zhu X (2014) NG2 cells (polydendrocytes) in brain physiology and repair. Front Neurosci 8:133. doi:10.3389/fnins.2014.00133
10. Warrington AE, Barbarese E, Pfeiffer SE (1993) Differential myelinogenic capacity of

specific developmental stages of the oligodendrocyte lineage upon transplantation into hypomyelinating hosts. J Neurosci Res 34(1): 1–13

11. Duncan ID, Paino C, Archer DR, Wood PM (1992) Functional capacities of transplanted cell-sorted adult oligodendrocytes. Dev Neurosci 14(2):114–122

12. Gard AL, Pfeiffer SE (1993) Glial cell mitogens bFGF and PDGF differentially regulate development of O4+GalC- oligodendrocyte progenitors. Dev Biol 159(2):618–630

13. Barres BA, Raff MC (1993) Proliferation of oligodendrocyte precursor cells depends on electrical activity in axons. Nature 361(6409): 258–260

14. Dugas JC, Emery B (2013) Purification and culture of oligodendrocyte lineage cells. Cold Spring Harb Protoc 2013(9):810–814. doi:10.1101/pdb.top074898

15. Goldman JE, Geier SS, Hirano M (1986) Differentiation of astrocytes and oligodendrocytes from germinal matrix cells in primary culture. J Neurosci 6(1):52–60

16. McCarthy KD, de Vellis J (1980) Preparation of separate astroglial and oligodendroglial cell cultures from rat cerebral tissue. J Cell Biol 85(3):890–902

17. Szuchet S, Yim SH (1984) Characterization of a subset of oligodendrocytes separated on the basis of selective adherence properties. J Neurosci Res 11(2):131–144

18. Chen Y, Balasubramaniyan V, Peng J, Hurlock EC, Tallquist M, Li J, Lu QR (2007) Isolation and culture of rat and mouse oligodendrocyte precursor cells. Nat Protoc 2(5):1044–1051. doi:10.1038/nprot.2007.149

19. Avellana-Adalid V, Nait-Oumesmar B, Lachapelle F, Baron-Van Evercooren A (1996) Expansion of rat oligodendrocyte progenitors into proliferative "oligospheres" that retain differentiation potential. J Neurosci Res 45(5):558–570

20. Zhang SC, Lipsitz D, Duncan ID (1998) Self-renewing canine oligodendroglial progenitor expanded as oligospheres. J Neurosci Res 54(2):181–190

21. Zhang SC, Lundberg C, Lipsitz D, O'Connor LT, Duncan ID (1998) Generation of oligodendroglial progenitors from neural stem cells. J Neurocytol 27(7):475–489

22. Li J, Lin JC, Wang H, Peterson JW, Furie BC, Furie B, Booth SL, Volpe JJ, Rosenberg PA (2003) Novel role of vitamin k in preventing oxidative injury to developing oligodendrocytes and neurons. J Neurosci 23(13):5816–5826

23. Raff MC, Miller RH, Noble M (1983) A glial progenitor cell that develops in vitro into an astrocyte or an oligodendrocyte depending on culture medium. Nature 303(5916):390–396

24. Dugas JC, Tai YC, Speed TP, Ngai J, Barres BA (2006) Functional genomic analysis of oligodendrocyte differentiation. J Neurosci 26(43):10967–10983

25. Emery B, Agalliu D, Cahoy JD, Watkins TA, Dugas JC, Mulinyawe SB, Ibrahim A, Ligon KL, Rowitch DH, Barres BA (2009) Myelin gene regulatory factor is a critical transcriptional regulator required for CNS myelination. Cell 138(1):172–185. doi:10.1016/j.cell.2009.04.031. S0092-8674(09)00456-5 [pii]

26. Weng Q, Chen Y, Wang H, Xu X, Yang B, He Q, Shou W, Higashi Y, van den Berghe V, Seuntjens E, Kernie SG, Bukshpun P, Sherr EH, Huylebroeck D, Lu QR (2012) Dual-mode modulation of Smad signaling by Smad-interacting protein Sip1 is required for myelination in the central nervous system. Neuron 73(4):713–728. doi:10.1016/j.neuron.2011.12.021. S0896-6273(12)00040-2 [pii]

27. Yu Y, Chen Y, Kim B, Wang H, Zhao C, He X, Liu L, Liu W, Wu LM, Mao M, Chan JR, Wu J, Lu QR (2013) Olig2 targets chromatin remodelers to enhancers to initiate oligodendrocyte differentiation. Cell 152(1-2):248–261. doi:10.1016/j.cell.2012.12.006. S0092-8674(12)01493-6 [pii]

28. Chan JR, Watkins TA, Cosgaya JM, Zhang C, Chen L, Reichardt LF, Shooter EM, Barres BA (2004) NGF controls axonal receptivity to myelination by Schwann cells or oligodendrocytes. Neuron 43(2):183–191

29. Watkins TA, Emery B, Mulinyawe S, Barres BA (2008) Distinct stages of myelination regulated by gamma-secretase and astrocytes in a rapidly myelinating CNS coculture system. Neuron 60(4):555–569. doi:10.1016/j.neuron.2008.09.011. S0896-6273(08)00759-9 [pii]

30. Zhang Y, Chen K, Sloan SA, Bennett ML, Scholze AR, O'Keeffe S, Phatnani HP, Guarnieri P, Caneda C, Ruderisch N, Deng S, Liddelow SA, Zhang C, Daneman R, Maniatis T, Barres BA, Wu JQ (2014) An RNA-sequencing transcriptome and splicing database of glia, neurons, and vascular cells of the cerebral cortex. J Neurosci 34(36):11929–11947. doi:10.1523/JNEUROSCI.1860-14.2014

Chapter 7

Schwann Cell Isolation and Culture Reveals the Plasticity of These Glia

David E. Weinstein

Abstract

In the chapter that follows we elaborate the methods for isolation of Schwann cells from adult as well as neonatal nerve and from a number of species, including humans. The ability to isolate, purify, and expand these cells has revealed their unusual plasticity: Schwann cells do not senesce, and therefore can be greatly expanded in culture, and once expanded, they can be induced to behave like uncommitted neural crest progenitor cells, able to give rise to neurons as well as more Schwann cells, depending upon the signals they receive. Finally, we have previously demonstrated that Schwann cells expanded in vitro can be used to engraft a biological bridge between distant axonal stumps, promoting axonal regeneration when none would otherwise occur.

Key words Schwann cell, Myelin, Axon, Plasticity, Nerve, Neural crest, Tissue culture, Phenotype

1 Introduction

During mammalian embryonic development the neural crest cells differentiate at the dorsal portion of the neural tube and then migrate throughout the developing organism, giving rise to a wide range of tissues, including the progenitors of the peripheral nervous system (PNS). In specific, the neural crest stem cells that migrate between the neural tube and the somites are exposed to a number of factors for varying lengths of time. The length of exposure, as well as the concentration and combination of growth factor exposure dictates the cell fate decisions that these cells will adopt. In specific, exposure to bone morphogenetic protein-2 (BMP-2) drives a neuronal phenotype, while exposure to glial growth factor-2 (GGF-2) for 48–96 h drives a Schwann cell phenotype, and a 24 h exposure to transforming growth factor beta-1 (TGFβ-1) drives a smooth muscle fate [1–3]. In contrast, the neural crest stem cells that migrate between the somites and the overlying ectoderm are instructed to a melanocyte fate [4, 5]. Alterations in any of these parameters can have profound effects on the organism [6].

Amit K. Srivastava et al. (eds.), *Stem Cell Technologies in Neuroscience*, Neuromethods, vol. 126,
DOI 10.1007/978-1-4939-7024-7_7, © Springer Science+Business Media LLC 2017

The neural crest stem cells that become committed neural cell lineages coalesce to give rise to the autonomic ganglia (sympathetic ganglia and parasympathetic ganglia), to dorsal root ganglia, and soon thereafter the new neurons begin to extend bundles of axonal projections to their target tissues [7]. During embryogenesis and into early postnatal life the Schwann cell progenitors continue to proliferate as they migrate along young axons, segregating them into progressively smaller bundles pinching off single axons, and establishing the Schwann cell–axon unit that is characteristic of the mature, myelinated nerve [8].

Unlike the central nervous system, in which adult homeostatic glia functions are split between the astrocyte and the oligodendrocyte, all PNS glial functions, including myelination, trophic and structural support and extracellular buffering, are performed by the mature Schwann cells (reviewed in [9–13]). The functional status of the adult nervous system is dependent upon the careful and balanced integration of these Schwann cell activities and their interactions with the associated axonal elements. Together, they allow for the coherent and synchronized firing of action potentials that underlie movement and sensory input. As an example, salutatory conduction down a myelinated nerve is dependent upon the axon firing action potentials and the associated Schwann cells buffering the extracellular matrix at the node of Ranvier, allowing for rapid repolarization. Alterations in Schwann cell function leads to profound deficits in PNS function [14, 15]. The dynamic nature of bidirectional signaling between axon and Schwann cell is exemplified by the constant ratio between axonal diameter and myelin thickness—called the g-ratio. Under normal conditions, the Schwann cell responds to increases in axonal diameter by increasing the turns of myelin around the axon, such that very small axons have only a few turns of myelin, while large caliber fibers can be wrapped with 50 of more layers of Schwann cell myelin [16]. Alterations in the g-ratio has profound effects of neuronal function, including slowing of nerve conductance in myelin hypomorphs [17], and increased conduction in myelin hypermorphs [18]. Birchmeier and Knave have shown that signaling between axonally expressed neuregulin-1 and its Schwann cell-expressed receptor ErbB3 regulate the initiation and extent of myelination [19]. These signals are integrated in the Schwann cell's nucleus, where transcriptional events control the onset and extent of myelin gene expression, which remains stable unless the nerve is damaged and axonal diameters change [20, 21].

The anatomy of the peripheral nerve rapidly changes following compression injury, when peripheral nerves undergo a stereotyped pattern of Wallerian degeneration, characterized by myelin decompaction and autophagocytosis of the myelin debris, recruitment of cells of the monocyte/macrophage lineage and axonal die-back (reviewed in detail in [22] and in [8]). With the loss of their axonal counterparts, the Schwann cells in the distal nerve undergo as round of proliferation, a phenomenon that was first reported 30 or more years ago [23] and confirmed again more recently [24]. Thereafter, the Schwann cells in the distal stump rapidly switch to a regenerative phenotype that includes the upregulated expression of c-*jun*, which in turn drives the expression of a range of trophic factors that promote axonal regrowth down the distal nerve stump [25]. We, and many others have shown that there is an absolute requirement for the presence of Schwann cell in the distal nerve for axonal reentry and for functional recovery [18, 26] and reviewed in [8]. Gaps in peripheral nerves beyond 3 cm will not regrow, even if the ends of a nerve are bridged [27]. In contrast, we have shown axons will transverse through a 6 cm or larger gap, to synapse on the appropriate downstream targets if, and only if, a donor blood vessel or other tubular bridge is prefilled with autologous Schwann cells prior toplacing it as an interpositional graft between the two nerve stumps [26]. To accomplish a nerve bridge a supply of autologous Schwann cells or Schwann cells of the same species is required. Thus it is necessary to isolate, enrich, and expand the Schwann cells.

Unlike almost any other mammalian cell types, Schwann cells do not senesce in culture [28], and thus can be repeatedly passaged and expanded. Moreover, once in culture, Schwann cells have a remarkable plasticity. Widera and colleagues have demonstrated that when cultured rat Schwann cells are grown under standard conditions, and then switched to serum-free media, they form neurospheres [29] that are highly reminiscent of the neurospheres generated from the subventricular zone in the brain [30, 31]. We have harvested Schwann cells from biopsy of adult human nerve, expanded them in culture and then switched them to a standard serum-free medium, whereupon the cells form spheres over the ensuing days, and processes begin to extend from the spheres (Fig. 1a through 1d). Staining these clusters with an antiserum to neuron-specific Tubulin III revealed extensive axonal growth both within the sphere and (1f) and extending beyond the sphere (1e). As the starting cells were taken from the mid-section of a cutaneous peripheral nerve, it was highly unlikely that any neuronal cells were included in the original material. Thus, these findings demonstrate the plasticity of the adult human Schwann cell.

Regardless of the desired use—either for therapeutic application, or to better understand their plasticity, the ability to growth these remarkable cells in culture is required. What follows is a description of methods for the harvesting, purification, and expansion of adult as well as neonatal Schwann cells in vitro.

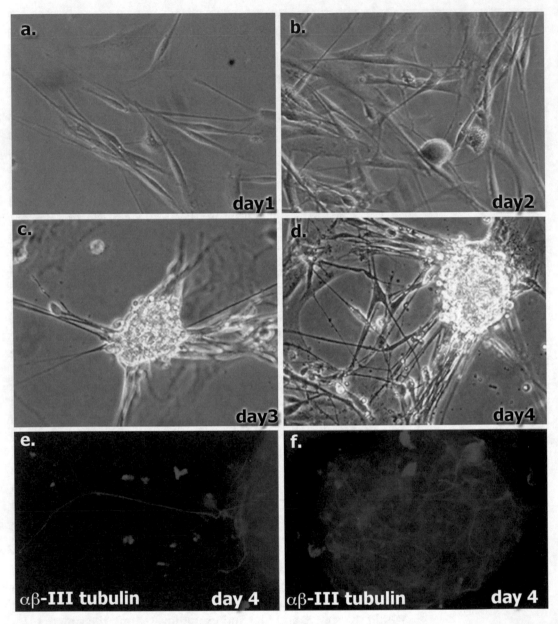

Fig. 1 Plasticity of adult Schwann cells is revealed in culture. Human Schwann cells were harvested from an adult surgical resection of breast tissue following mammoplasty. The peripheral nerves were identified, harvested and expanded as described above. After purification of Schwann cells, the cells were plated at clonal density in Terasaki wells and expanded as described above. After the cells reached 50% confluence, they were switched to Sato's serum-free medium (1:1 Dulbecco's MEM and Ham's F-12 supplemented with ITS, progesterone, T3, 0.1%BSA). The times indicate the number of days after switch to Sato's medium (days 1 through 4). On the first day after the switch, the majority of the cultured cells still have an adherent, bipolar phenotype consistent with cultured Schwann cells (panel **a**). As early as 2 days after switch, the cells begin to form spheres (panel **b**), and as the spheres grow in size the cells within the sphere begin to send out processes (**c** and **d**). Staining of these cultures 4 days after medium switch (**e** and **f**) reveals that a subset of cells have undergone a fate switch from Schwann cell to neuron, as they are now expressing the neuronal marker βIII tubulin, and extending long axons

2 Materials

Solutions:

DMEM + 10% FCS	DMEM supplemented with 10% heat inactivated FCS, 1% DMEM nonessential amino acids (11140-019, Gibco) and penicillin/streptomycin/glutamine (#400-110, Gemini Bioproducts). Store at 4 °C.
Ham's F10 + antibiotics	Ham's F10 supplemented with penicillin/streptomycin/glutamine (#400-110, Gemini Bioproducts)
DNase	7.5 mg DNase (DP DNase, Worthington Biochem) in 15 ml DMEM. *DO NOT pH!* Store at −20 °C.
PBS (10×)	80 g NaCl, 2 g KCl, 21.6 g Na_2HPO_4, 2 g KH_2HPO_4, pH to 7.4, add H_2O to 1 L. Store at room temperature.
Trypsin	150 mg trypsin (TRL trypsin, Worthington Biochem), 15 mg DNase (DP DNase, Worthington Biochem), 15 ml PBS supplemented with Mg^{++} (0.5 mg $MgSO_4$= >100 ml PBS), 90 μl 1 N NaOH. Store at −20 °C.
AraC (1000×)	27.9 mg araC (Sigma C-6645) into 10 ml of ddH_20. Store frozen at −20 °C.
Trypan Blue Solution	T8154 Sigma-Aldrich
Bovine Pituitary Extract	Sigma P-1167
Collagenase, 10 mg/ml	Sigma C 9891
Glial Growth Factor2	MyBiosource catalog # MBS553231
Trypsin/EDTA	Gibco, # 25300-062

3 Methods

3.1 Isolation of Schwann Cells from Adult Nerve

3.1.1 Harvesting Nerve from Adult Rodent or Rabbit

The nerves of the lower limb are the ideal source, as they are the largest, and most readily available. However, if the lower limb is a problem, the upper limb nerves, such as the brachial, ulnar, and radial nerves can be used. If the animals are to be used for additional experimental steps all procedures must be carried out using good surgical techniques and Listerian antisepsis. Animals should be anesthetized with a combination of ketamine (75 mg/kg IM) and acepromazine (2.0 mg/kg IM), followed by buprenorphine (0.05 mg/kg IV) preoperatively. Anesthesia should then be maintained with ketamine (20 mg/kg IV) and acepromazine (2.0 mg/kg). To access the nerve thigh of the needs to be shaved and then prepped using a Betadine® solution followed by an alcohol wash. Approximately 3 cc of 2% lidocaine with epinepherine is to be infiltrated into the sciatic notch and an additional 3 cc used for subcutaneous injection along the nerve pathway. Using sterile surgical technique, an incision is made from the sciatic notch to the femoral head. The fascia of the biceps femoris and vastus lateralis separated, exposing the sciatic, common peroneal, and tibialis nerves.

The length of the exposed nerve then transected both proximally and distally using a #15 blade and excised. The harvested nerve is then placed in a tube filled with ice-cold PBS, and the Schwann cells isolated, as described below. The leg is then surgically closed, and the animal placed in the left lateral position on a warming pad until it awakens. Buprenorphine (0.05 mg/kg IV) can be given for 3 days postoperatively for pain management.

If the animals are not to be used for additional experimental steps, additional nerves can be harvested to increase the Schwann cell yield. When the nerve harvest is complete a lethal dose of ketamine and acepromazine, followed by barbituric acid can be administered intravenously, and more extensive sections of nerve, including the full length of the sciatic nerve, can be harvested. The nerves should then be dissected free of surrounding tissue and placed into a fresh petri dish filled with sterile ice-cold PBS or 0.9% NaCl and washed several times with sterile ice-cold PBS or 0.9% NaCl to remove excess blood or non-neural tissue contaminants.

3.1.2 Harvesting Human Nerve

Peripheral nerve can be harvested from a number of human sources, including surgical resections, such as breast reduction tissue, early post mortem tissue and abortus. The source depends on availability, but the techniques employed are virtually identical. Because human peripheral nerve myelination mostly occurs *in utero*, virtually all tissue samples available for Schwann cell isolation must be handled like adult tissue.

The tissue should be obtained from the operating suite or the autopsy suite as soon as it is available after specimens have been taken for clinical needs. The tissue should be placed in ice-cold saline or ice-cold PBS in a tube or other suitable, sealed container, and the container should be transported to the laboratory on ice. Researchers working with this material must be aware that human tissue should be considered as a pathogen risk. Gloves and gowns should be used at all times when handling the material. Upon arrival in the laboratory, the specimen should be transferred to a large petri dish filled with sterile ice-cold PBS or 0.9% NaCl in a tissue culture hood. The tissue should be washed several times with sterile ice-cold PBS or 0.9% NaCl to remove excess blood and any surface contaminants. The tissue is then blunt dissected until one or more sizeable peripheral nerves are identified. The nerves should then be dissected free of surrounding tissue and placed into a fresh petri dish filled with sterile ice-cold PBS or 0.9% NaCl and washed several times with sterile ice-cold PBS or 0.9% NaCl to remove excess blood or non-neural tissue contaminants.

3.2 Isolating Schwann Cells from the Adult Nerve and Human Nerve of Any Age

The thick, vascular ectoneurium, which is rich in fibroblasts, vascular smooth muscle cells, and endothelial cells, must be removed prior to isolation of the Schwann cells. This is best accomplished by grasping an axonal fascicle, which is heavily myelinated, and thus white, with one pair of forceps. With a second forceps, grasp the ectoneurial membrane and pull it down the length of the axons. This is analogous to pulling a stocking down a leg, as the ectoneurium will evert as it is pulled down the length of the nerve. The ectoneurium is quite fibrous and tough in the adult, so do not hesitate to use force to pull it down as it is turning inside-out. The use of a rat-toothed forceps eases the grasping of the ectoneurium. Discard the thus removed ectoneurial tissue. Move the exposed endoneurium to a fresh petri dish and cover in ice-cold buffer.

While still in the petri dish cut the nerve into 3–5 mm pieces, using a sterile razor blade, or scalpel blade. This is best accomplished by placing the cutting edge against the nerve and applying downward pressure. We strongly advise against sawing through the tissue with a back-and-forth motion, as this will stretch and distort the tissue without actually cutting it into pieces. This procedure, as with the others, must be accomplished with sterile technique. The pieces of nerve are transferred to a 15 ml conical tube that has 2 ml of Ham's F10 supplemented with antibiotics and 1 ml of collagenase. If necessary, More Ham's F10 supplemented with antibiotics plus collagenase can be added to cover the tissue completely. The tube, with the lid sufficiently loose to allow gas exchange, is then placed into a rack in a 37 °C humidified incubator, buffered with 7% CO_2.

Following a 2.5-h digestion with collagenase, the tube of tissue is moved back to the tissue culture hood, where 1 ml of trypsin solution (or more if needed to fully cover the tissue) (recipe follows) is added for 7 min, and the tissue is then centrifuged for 3 min at 1200 rpm (approximately 250 g, depending on the rotor size). It is critical that the trypsin solution be made as specified, with reagents from specified vendors or there will be a very poor yield of Schwann cells.

Following trypsin treatment, the supernatant is gently removed with a Pasteur pipette and 2 ml of DNase solution is added. The tissue is then taken through graded needles. Beginning with the 18 gauge needle, draw up the tissue and DNase solution and expel it back into the tube. Do not draw up air, as this will create bubbles and significantly reduce yield. Repeat this procedure for a total of 15 times, after which change to a 20 gauge needle and then the 23 gauge needle repeating the mechanical disruption 15 times with each size needle. The needle passage is intended to disrupt the tissue, and generate a single cell suspension. After the final needle, add 10 ml of DMEM + 10% FCS w and the cells are spun for 10 min, at 1200 rpm, preferably at 4 °C. After the spin step remove the supernatant and resuspend the cells in 10 ml of DMEM + 10% FCS.

The cell suspension is then plated onto a 10 cm Primaria® tissue culture plate, and cultured in a 37 °C humidified incubator, buffered with 7% CO_2. Primaria® plates are used to minimize fibroblast adhesion and growth as the Schwann cells are expanded. It should be anticipated that there will be a large amount of axonal and myelin debris after this overnight culture step, which will be removed the following morning by rinsing the plate and replacing the DMEM + 10% FCS with DMEM + 10% FCS. Culture the cells overnight in a humidified incubator at 37 °C/7%CO$_2$. Following overnight culture, the medium is changed to DMEM + 10% FCS, supplemented with 100 μM of araC and cultured for an additional 48 h to remove the last contaminating fibroblasts.

After araC treatment the plates are gently rinsed with DMEM warmed to 37 °C supplemented ×3 and the medium is replace with DMEM + 10% FCS supplemented with 2 μM forskolin and either recombinant glial growth factor-2 at 50 ng/ml, or bovine pituitary extract at 10 μg/ml. Replace the cells in the incubator and grow them until they reach 90% confluence.

3.3 Final Schwann Cell-Enrichment Steps

This is an optional step intended to remove any last contaminating fibroblasts through antibody/compliment-mediated cytolysis. The target antigen is Thy 1.1 or 1.2, or CD90 in humans, expressed on fibroblasts, but not Schwann cells. Depending of which species is used of Schwann cell is being isolated, the appropriate antibody must be used. Check with antibody manufacturers for the concentrations and conditions needed for antibody/compliment-mediated cytolysis.

When the Schwann cell enriched cultures reach confluence, trypsinize the cells with trypsin–EDTA. Wash the cells with PBS ×3, remove the last wash and add 1 ml of trypsin/30 mm of tissue culture surface area. Make frequent microscopic observations. As soon as the cells begin to loosen from the plate, tap the plate firmly against a hard surface to facilitate the lifting of the cells. As soon as the majority of cells have detached from the surface add 3× volume of DMEM + 10% FCS, and centrifuge at 1200 rpm for 5 min. Resuspend the cell pellet in 5 ml of DMEM +10% FCS and centrifuge at ~150 × g for 5 min. Resuspend the pellet in 1 ml of antibody containing solution, as recommended by the manufacturer for cytolysis. Incubate the cells on ice for 1–2 h, and spin in a refrigerated centrifuge at ~150 × g for 5 min, then resuspend the cell pellet in 3 ml of DMEM and add 1 ml of rabbit or guinea pig complement. Incubate at 37 °C × 1 h, followed by dilution to full volume of the tube with DMEM wand centrifuge the cells at ~150 × g for 5 min, resuspend the pellet in DMEM and centrifuge at ~150 × g for 5 min. Repeat the wash step two times. Resuspend the cells and count them in a hemocytometer, using trypan blue to determine cell viability and efficacy of antibody mediated cytolysis. The remaining cells are essentially pure Schwann cells cultures, and can now be grown, used or passaged. Plate cells at 5 × 10^4/plate on Primaria® plates in DMEM + 10% FCS supplemented with 2 μM forskolin and either recombinant glial growth factor2 (GGF2) at 50 ng/ml, or bovine pituitary extract at 10 μg/ml. The forskolin upregulates cAMP in the Schwann cells which then synergizes with GGF/BPE to promote Schwann cell proliferation [32]. Passage the cells at confluence, splitting them 1:4.

3.4 Isolating Schwann Cells from Neonatal Nerve

The presence of myelin in adult nerve decreases the yield of viable cells. To overcome this obstacle, the following protocol describes the isolation of Schwann cells from neonatal animals, before the onset of myelination [20]. This protocol assumes that the sciatic nerve will be the source of Schwann cells because of its size and easy accessibility. However, any peripheral nerve, or combination of nerves, can be used as a starting source. Our protocol is designed for an entire litter or more of postnatal day (P) 0 to P4 pups, but can be carried out with as little as a single nerve.

Because young animals are less susceptible to common anesthetics, the entire litter is cooled on ice prior to initiation of the harvest. Once an animal has stopped moving, decapitate it with the large scissors, dropping the head into a collection bag and pin the body to a dissecting board, ventral side down. Remove the dorsal skin of the legs and lower torso, and separate the muscles of the lower back and upper thigh by blunt dissection to reveal the sciatic nerve. Using either a fresh scalpel blade or a sharp iris scissor cut the nerve at the knee at spinal column. Grasp the nerve with a smooth-bladed forceps and gently lift it away from the body, cutting away any adherent tissue as you lift, taking care at all times not to stretch or pull the nerve, and transfer to a 5 ml collection tube filled with ice-cold DMEM or PBS. Repeat this process on the contralateral nerve, and on all the pups in the litter, collecting all the nerves in a single, 5 ml tube on ice.

3.5 Schwann Cell Enrichment from Neonatal Nerve

Gently remove the buffer from the tube and immediately add 1 ml of the DMEM plus 300 μl of collagenase solution, making sure that the fluid is in contact with all of the tissue in the tube, cap the tube and transfer it to a 37 °C water bath for 30 min. Then remove the DMEM/collagenase and immediately add 1 ml of trypsin solution, making sure that the fluid is in contact with all of the tissue in the tube. Replace the cap and transfer it to a 37 °C water bath for 5 min, after which remove the trypsin with a Pasteur pipette and bulb, taking care not to remove the tissue, which is now quite friable. Carefully wash the tissue three times with 1–2 ml of ice-cold PBS by adding ~2 ml of wash buffer, allow the tissue to settle to the bottom of the tube and then remove the buffer with a Pasteur pipette and bulb, making sure that the pellet is undisturbed.

After removing the last wash solution, add 1 ml of DNase solution. From this step forward, the isolation of Schwann cells is identical in adult, neonatal, and human nerve of any age, and is explained above.

4 Notes

It is critical that the trypsin used for tissue digestion is the *TRL* trypsin powder from Worthington Biochem, and the DNase powder is the *DP DNase* powder from Worthington Biochem, otherwise there will be extremely poor cell yields with little to no viability.

References

1. Shah NM, Anderson DJ (1997) Integration of multiple instructive cues by neural crest stem cells reveals cell-intrinsic biases in relative growth factor responsiveness. Proc Natl Acad Sci U S A 94(21):11369–11374

2. Shah NM, Groves AK, Anderson DJ (1996) Alternative neural crest cell fates are instructively promoted by TGFbeta superfamily members. Cell 85(3):331–343

3. Shah NM, Marchionni MA, Isaacs I, Stroobant P, Anderson DJ (1994) Glial growth factor restricts mammalian neural crest stem cells to a glial fate. Cell 77(3):349–360

4. Creuzet S, Couly G, Le Douarin NM (2005) Patterning the neural crest derivatives during development of the vertebrate head: insights from avian studies. J Anat 207(5):447–459. doi:10.1111/j.1469-7580.2005.00485.x

5. Le Douarin NM, Creuzet S, Couly G, Dupin E (2004) Neural crest cell plasticity and its limits. Development 131(19):4637–4650. doi:10.1242/dev.01350

6. Joseph NM, Mukouyama YS, Mosher JT, Jaegle M, Crone SA, Dormand EL, Lee KF, Meijer D, Anderson DJ, Morrison SJ (2004) Neural crest stem cells undergo multilineage differentiation in developing peripheral nerves to generate endoneurial fibroblasts in addition to Schwann cells. Development 131(22):5599–5612. doi:10.1242/dev.01429

7. Webster DF (ed) (1984) Development of peripheral nerve fibers. In: Peripheral neuropathy. W. B. Saunders, Philadelphia

8. Weinstein DE (1999) The role of Schwann cells in neural regeneration. Neuroscientist 5(4):208–216

9. Suter U, Martini R (2005) Myelination. In: Dyck PJ, Thomas PK (eds) Peripheral neuropathy, vol 1. W. B. Saunders Company, Philadelphia, pp 411–431

10. Weinstein DE (1998) Myelination: coordinated regulation of many molecular elements. Ment Retard Dev Disabil Res Rev 4:179–186

11. Bunge MB (1993) Schwann cell regulation of extracellular matrix biosynthesis and assembly. In: Dyck PJ, Thomas PK (eds) Peripheral neuropathy, vol 1. W. B. Saunders Company, Philadelphia, pp 299–316

12. Chiu SY (2005) Channel function in mammalian axons and support cells. In: Dyck PJ, Thomas PK (eds) Peripheral neuropathy, vol 1, 4th edn. W. B. Saunders Company, Philadelphia, pp 95–112

13. Windebank J, McDonald ES (2005) Neurotrophic factors in the peripheral nervous system. In: Dyck PJ, Thomas PK (eds) Peripheral neuropathy, vol 1, 4th edn. W. B. Saunders Company, Philadelphia, pp 377–386

14. Rosenbluth J (1979) Aberrant axon-Schwann cell junctions in dystrophic mouse nerves. J Neurocytol 8(5):655–672

15. Stirling CA (1975) Abnormalities in Schwann cell sheaths in spinal nerve roots of dystrophic mice. J Anat 119(1):169–180

16. Waxman SG, Bennett MV (1972) Relative conduction velocities of small myelinated and non-myelinated fibres in the central nervous system. Nat New Biol 238(85):217–219

17. Baloh RH, Strickland A, Ryu E, Le N, Fahrner T, Yang M, Nagarajan R, Milbrandt J (2009) Congenital hypomyelinating neuropathy with lethal conduction failure in mice carrying the Egr2 I268N mutation. J Neurosci 29(8):2312–2321. doi:10.1523/JNEUROSCI.2168-08.2009

18. Bieri PL, Arezzo JC, Weinstein DE (1997) Abnormal nerve conduction studies in mice expressing a mutant form of the POU transcription factor SCIP. J Neurosci Res 50(5):821–828

19. Birchmeier C, Nave KA (2008) Neuregulin-1, a key axonal signal that drives Schwann cell growth and differentiation. Glia 56(14):1491–1497. doi:10.1002/glia.20753

20. Weinstein DE, Burrola PG, Lemke G (1995) Premature Schwann cell differentiation and hypermyelination in mice expressing a targeted antagonist of the POU transcription factor SCIP. Mol Cell Neurosci 6(3):212–229

21. Wu R, Jurek M, Sundarababu S, Weinstein DE (2001) The POU gene brn-5 is induced by neuregulin and is restricted to myelinating Schwann cells. Mol Cell Neurosci 17(4):683–695

22. Griffin JW, George EB, Chaudhry V (1996) Wallerian degeneration in peripheral nerve disease. Baillieres Clin Neurol 5(1):65–75

23. Salzer JL, Williams AK, Glaser L, Bunge RP (1980) Studies of Schwann cell proliferation. II. Characterization of the stimulation and specificity of the response to a neurite membrane fraction. J Cell Biol 84(3):753–766

24. Yang DP, Zhang DP, Mak KS, Bonder DE, Pomeroy SL, Kim HA (2008) Schwann cell proliferation during Wallerian degeneration is not necessary for regeneration and remyelination of the peripheral nerves: axon-dependent removal of newly generated Schwann cells by apoptosis. Mol Cell Neurosci 38(1):80–88. doi:10.1016/j.mcn.2008.01.017

25. Arthur-Farraj PJ, Latouche M, Wilton DK, Quintes S, Chabrol E, Banerjee A, Woodhoo A, Jenkins B, Rahman M, Turmaine M, Wicher GK, Mitter R, Greensmith L, Behrens A, Raivich G, Mirsky R, Jessen KR (2012) c-Jun reprograms Schwann cells of injured nerves to generate a repair cell essential for regeneration. Neuron 75(4):633–647. doi:10.1016/j.neuron.2012.06.021

26. Strauch B, Rodriguez DM, Diaz J, Yu HL, Kaplan G, Weinstein DE (2001) Autologous Schwann cells drive regeneration through a 6-cm autogenous venous nerve conduit. J Reconstr Microsurg 17(8):589–595 . doi:10.1055/s-2001-18812discussion 596-587

27. Strauch B, Ferder M, Lovelle-Allen S, Moore K, Kim DJ, Llena J (1996) Determining the maximal length of a vein conduit used as an interposition graft for nerve regeneration. J Reconstr Microsurg 12(8):521–527

28. Mathon NF, Malcolm DS, Harrisingh MC, Cheng L, Lloyd AC (2001) Lack of replicative senescence in normal rodent glia. Science 291(5505):872–875. doi:10.1126/science.1056782

29. Martin I, Nguyen TD, Krell V, Greiner JF, Muller J, Hauser S, Heimann P, Widera D (2012) Generation of Schwann cell-derived multipotent neurospheres isolated from intact sciatic nerve. Stem Cell Rev 8(4):1178–1187. doi:10.1007/s12015-012-9387-2

30. Walker TL, White A, Black DM, Wallace RH, Sah P, Bartlett PF (2008) Latent stem and progenitor cells in the hippocampus are activated by neural excitation. J Neurosci 28(20):5240–5247. doi:10.1523/JNEUROSCI.0344-08.2008

31. Young KM, Merson TD, Sotthibundhu A, Coulson EJ, Bartlett PF (2007) p75 neurotrophin receptor expression defines a population of BDNF-responsive neurogenic precursor cells. J Neurosci 27(19):5146–5155. doi:10.1523/JNEUROSCI.0654-07.2007

32. Rahmatullah M, Schroering A, Rothblum K, Stahl RC, Urban B, Carey DJ (1998) Synergistic regulation of Schwann cell proliferation by heregulin and forskolin. Mol Cell Biol 18(11):6245–6252

Chapter 8

Generation of Cerebral Organoids Derived from Human Pluripotent Stem Cells

Mark E. Hester and Alexis B. Hood

Abstract

Human cortical brain development is a tightly controlled and highly orchestrated process composed of neural progenitor cell (NPC) proliferation, migration, differentiation, and maturation. Recent advances in cerebral organoid technology have provided a means to model these complex cellular mechanisms to advance our understanding of normal human brain development and to provide molecular insight into the pathogenesis of brain disease. Cerebral organoids, which are generated from human pluripotent stem cells, are composed of three-dimensional neural tissue. This tissue can self-organize to form discrete regions of the human brain that includes cerebral cortex, choroid plexus, and others, if appropriate differentiation cues are present. Indeed, cerebral organoids provide an invaluable resource to study human-specific aspects of corticogenesis in vitro, such as investigating the function of outer radial glia (oRG), and other complex features of the human cerebral cortex. Here, we provide an overview of several methodologies to generate human cerebral organoids derived from human pluripotent stem cells including modifications from our laboratory. In addition, we highlight the advantages and current challenges associated with using cerebral organoids as an in vitro system to model human cortical brain development and disease.

Key words Cerebral organoids, Induced pluripotent stem cell, Human brain development, Neuronal differentiation, Cerebral cortex, 3D culture system, Neurogenesis, Neurodevelopment, Outer radial glia, Progenitor zone organization

1 Introduction

Encompassing greater than three-quarters of the human brain and functioning as the major control center for the central nervous system (CNS), the cerebral cortex is one of the most complex and intricate structures of the human body. The complexity of the cerebral cortex is the result of a highly coordinated developmental program that initiates from a population of multipotent progenitor cells residing predominately in the ventricular zone (VZ) [1–4]. These cortical progenitors are then temporally specified into distinct neural subtypes within a stereotypical six-layered structure termed the cortical plate by migrating from the inner most cortical layer to the outer most layer during development [2]. Historically,

Amit K. Srivastava et al. (eds.), *Stem Cell Technologies in Neuroscience*, Neuromethods, vol. 126,
DOI 10.1007/978-1-4939-7024-7_8, © Springer Science+Business Media LLC 2017

the research community has relied predominately on rodent models to elucidate the cellular mechanisms that control cortico-genesis. Recently, innovations in stem cell technologies have fueled development of numerous stem cell-based systems to model human cerebral corticogenesis [5–12].

Here we provide an overview of several current methods to generate human brain organoids derived from pluripotent stem cells and we also incorporate additional methodologies to stream-line the process [7, 8, 10, 11, 13]. The first report to generate 3D neural tissue in vitro that recapitulates hallmarks of the developing human cerebral cortex, including cortical progenitor ventricular zones, stratification of the cortical plate, and structures that express choroid plexus markers was published by Lancaster et al. in 2013. Their method built upon previous studies that defined optimal conditions for differentiating human PSC-derived embryoid bod-ies into cortical neural lineages [14–19]. The authors then coupled this knowledge with a protocol whereby neuroectodermal bodies were encapsulated into Matrigel droplets, which provided a struc-tural matrix for promoting the expansion of neuroepithelial buds. Encapsulated neuroectodermal bodies are then cultured in a neural differentiation medium lacking vitamin A to promote expansion of neuroepithelium, and are allowed to grow over a period of 4 days. After this time period, cerebral organoids are then transferred into neural differentiation medium containing vitamin A and cultured with agitation using a spinning bioreactor. This technique allows for efficient nutrient diffusion and subsequent expansion, as well as development of cerebral brain tissue. Remarkably, cerebral organ-oids generated by this method can develop up to 4 mm in diameter within a time period of 2 months [11, 20].

Since the initial conception of cerebral organoid technology, alternative approaches and innovations to this protocol have been recently developed [8, 12]. Specifically, Qian et al. developed a protocol to generate brain-region-specific cerebral organoids, including midbrain and hypothalamic organoids within a custom-ized and miniaturized spinning bioreactor [8]. In addition, Lindborg et al. utilized hydrogel as an encapsulating material and cultured organoids in a defined cell culture medium to yield organ-oids up to 3 mm in size within 28 days [12].

A key advantage of these three-dimensional tissue culture sys-tems is that they have been demonstrated to recapitulate in vivo events more precisely, which allows for greater physiological rele-vance [21]. In conjunction with patient-specific iPSCs, cerebral organoid models not only hold immeasurable potential for study-ing the pathogenic mechanisms causing human brain diseases such as microcephaly and lissencephaly, but also could serve as a screen-ing tool to measure the toxicological effects of chemical and biological exposure on human brain development [22]. Importantly, these models provide a unique tool to measure molecular aspects of corticogenesis that is observed only in humans such as the relative abundance of ORGs, which are hypothesized to be responsible for the larger size of the human cerebral cortex compared to other mammalian species [1–3].

Although cerebral organoid technology holds tremendous potential and promise for deciphering mechanisms of human brain development and disease, there are several current limitations and challenges that need to be overcome. First, considerable variability can exist between different preparations of cerebral organoids and even within a single preparation, which poses a challenge to consistently assay, and compare cellular phenotypes between control and experimental groups [11, 21]. The development and use of validated markers, controls, and reporter systems that relay key developmental time points and cell fate identity would be highly beneficial to partially overcome this obstacle. In addition, live cell imaging analysis has proven to be difficult in thick 3D cultures due to ineffective light scattering within deep layers [21]. This type of analysis may also pose a challenge when imaging cerebral organoids, which can grow up to several millimeters in diameter. Apart from these technical difficulties, cerebral organoids are not completely representative of all cell types that contribute to corticogenesis, such as vascular and glial cells [11, 20]. Further improvements and innovations to this technology will undoubtedly uncover additional insight into their role and function during cortical development.

2 Materials

2.1 Supplies/Plasticware

1. Extra deep dish 10 cm² plates (ThermoFisher; Cat. No. 08-757-28).
2. Transfer pipets (ThermoFisher; Cat. No. 13-711-7M).
3. StemPro® EZPassage™ passage tool (ThermoFisher; Cat. No. A18945).
4. Cell Culture 100 mm dish (FisherSci; Cat. No. 08-772-23).
5. Fisherbrand razor blades (FisherSci; Cat. No. 12-640).
6. Parafilm (4 in. × 250 feet) (FisherSci; Cat. No. 13-374-12).
7. Fisherbrand™ Premium Microcentrifuge Tubes: 1.5 ml (FisherSci; Cat. No. 05-408-129).
8. 250 ml, CA filter (FisherSci; Cat. No. 09-761-1).

2.2 Cells, Media, and Supplements

1. Episomal iPSC Line (ThermoFisher; Cat. No. A18945).
2. Essential 8™ Flex Medium Kit (ThermoFisher; Cat. No. A2858501).
3. D-PBS (6× 1000 ml) (ThermoFisher; Cat. No. 14190235).
4. EDTA disodium 0.5 M (ThermoFisher; Cat. No. AM9260G).

5. GlutaMAX (ThermoFisher; Cat. No. 35050-061).

6. Nonessential amino acid solution (ThermoFisher; Cat. No. 11140-050).

7. KO serum replacement (ThermoFisher; Cat. No. 10828-028).

8. Antibiotic-Antimycotic (100×), liquid (ThermoFisher; Cat. No. 15240-062).

9. N2 supplement (ThermoFisher; Cat. No. 17502-048).

10. DMEM F12 (ThermoFisher; Cat. No. 11330-057).

11. Insulin solution (Sigma; Cat. No. I9278-5ML).

12. Heparin (Sigma; Cat. No H3419-10KU).

13. Vitronectin (ThermoFisher; Cat. No. A14700).

14. D-PBS (6× 1000 ml) (ThermoFisher; Cat. No. 14190235).

15. SM1 supplement minus vitamin A (Stemcell Technologies; Cat. No. 5731).

16. SM1 (STEMCELL Technologies; Cat. No. 5711).

17. ROCK inhibitor: Stemolecule™ Y27632 Stemgent (Stemgent; Cat. No. 04-0012-02).

18. BrainPhys Media (STEMCELL Technologies; Cat. No. 5790).

19. Growth factor reduced-Matrigel (ThermoFisher; Cat. No. CB-40230).

20. Fetal bovine serum, embryonic stem cell-qualified (ThermoFisher; Cat. No. 16141079).

21. 2-Mercaptoethanol (Merck, Cat. No. 8057400005).

22. Sterile Water.

23. Spray bottle containing 70% (vol/vol) ethanol.

24. Trypan blue (Bio-Rad; Cat. No. 145-0021).

2.3 Equipment

1. Pipetman P12X10L, 0.5–10 µl (Gilson; Cat. No. FA10014).

2. Pipetman P12X300L, 20–300 µl (Gilson; Cat. No. FA10016).

3. Pipetman P2L, 0.2–2 µl (Gilson; Cat. No. FA10001M).

4. MacroMan Pipette Aid (Gilson; Cat. No. F110756).

5. DF300ST Tipack, Filter tips (Gilson; Cat. No. F171603).

6. DFL10ST Tipack, Filter tips (Gilson; Cat. No. F171203).

7. DF1000ST Tipack, Filter tips (Gilson; Cat. No. F171703).

8. CO_2/O_2 tissue culture incubator (Panasonic; Cat. No. MCO-19M UV-PA).

9. Class 2 biological safety cabinet (NUAIRE; Cat. No. NU-425-400).

10. Conical tubes, 15 ml (Fisher Scientific, Cat. No. 05-527-90).

11. Water bath, 37 °C (Fisher Scientific, Isotemp water bath).

12. CO_2 Resistant Orbital Shaker (ThermoFisher; Cat. No. 88-881-103).

13. Transfer pipets (Fisher; Cat. No. 13-711-7M).

14. Zeiss Discovery Stereomicroscope (Carl Zeiss Microscopy, LLC).

15. CA filter, 500 ml, (FisherSci; Cat. No. 09-761-5).

16. 5 ml pipettes, Corning No.:4487 (FisherSci; Cat. No. 07-200-573).

17. 10 ml pipettes, Corning No.:4488 (FisherSci; Cat. No. 07-200-574).

18. 25 ml pipettes, Corning N 4489 (FisherSci; Cat. No. 07-200-575).

19. 5 ml aspirator pipettes (FisherSci; Cat. No. 13-675-10CC).

2.4 Media Preparation

1. EB Induction Medium (EIM): Combine 225 ml of DMEM-F12 with 5 ml of GlutaMAX, 5 ml of MEM-NEAA, 15 ml of hPSC-quality FBS, 50 ml of Knockout serum replacement, 225 ml of Essential 8™ complete medium, and 3.5 μl of 2-mercaptoethanol. Filter-sterilize medium using a 0.2-μm filter that is vacuum driven. Store for no longer than 2 weeks between 2 and 8 °C.

2. Neuroectoderm Induction Medium (NIM): Combine 500 ml of DMEM-F12 with 5 ml of N2 supplement, 5 ml GlutaMAX supplement, 5 ml MEM-NEAA, and 0.5 mg of Heparin (final concentration 1 μg/ml). Filter-sterilize medium using a 0.2-μm filter that is vacuum driven. Store for no longer than 2 weeks between 2 and 8 °C.

3. Cerebral Organoid Expansion Medium (COEM): Combine 250 ml of BrainPhysTM medium, 1.25 ml of N2 supplement, 2.5 ml SM1 supplement (lacking vitamin A), 1.25 ml of MEM-NEAA, 2.5 ml of GlutaMAX supplement, 62.5 μl of insulin, and 2.5 ml of penicillin–streptomycin. Add 87.5 μl of a 1:100 dilution of 2-mercaptoethanol. Filter-sterilize medium using a 0.2-μm filter that is vacuum driven. Store for no longer than 2 weeks between 2 and 8 °C.

4. Cerebral Organoid Growth and Differentiation Medium (COGDM): Combine 250 ml of BrainPhys™ medium, 1.25 ml of N2 supplement, 2.5 ml SM1 supplement (containing Vitamin A), 1.25 ml of MEM-NEAA, 2.5 ml of GlutaMAX supplement, 62.5 μl of insulin, and 2.5 ml of penicillin–streptomycin. Add 87.5 μl of a 1:100 dilution of 2-mercaptoethanol. Filter-sterilize medium using a 0.2-μm filter that is vacuum driven. Store for no longer than 2 weeks between 2 and 8 °C.

2.5 Preparation of Vitronectin-Coated Plate

Thaw a vial of vitronectin truncated recombinant human protein (VTN-N) from the −80 °C storage and prepare a 1:100 dilution in PBS (5 μg/ml). Pipet 6 ml of solution into a 10 cm² tissue culture dish to generate a final concentration of vitronectin at 0.5 μg/cm². Incubate the coated tissue culture plates for 1 h at room temperature. Large batches of coated plates that have been sealed with parafilm can be prepared and frozen at −20 °C storage for up to 1 month.

2.6 Preparation of Matrigel Aliquots

Thaw a 5 ml bottle of growth factor reduced-Matrigel on ice overnight at 4 °C. The next day before aliquoting, cool several 1 ml pipettes and ten 1.5 ml microfuge tubes at −20 °C for 15 min to prevent Matrigel from solidifying when pipetting. Using sterile conditions and working in a class 2 biological safety hood, pipette 500 μl of Matrigel into each tube and store at −80 °C for up to 1 year. Do not thaw and refreeze Matrigel aliquots as this will reduce the quality of the material.

3 Methods

3.1 Generation and Differentiation of Embryoid Bodies (EBs)

1. Culture hiPSCs in feeder-free conditions using a vitronectin-coated 10 cm² tissue culture dish and expand cells to 80% confluency or when colonies approach borders of neighboring colonies (*see* Fig. 1b). Closely monitor hiPSC cultures as continued growth of hiPSCs will result in spontaneous differentiation and will not yield optimal cerebral organoid formation (*see* **Note 1**).

2. Upon reaching 80% confluency, remove medium from hiPSCs and replace with 5 ml of fresh EIM. Use a StemPro® EZPassage™ tool to manually produce uniform patches (~300 μm²) of hiPSCs. With firm pressure, roll tool in one direction of the plate. Perform this procedure multiple times to ensure the whole plate has been scored in one direction. Next, rotate the plate 90° and roll the tool again over multiple passes to produce square grids of iPSC patches (*see* Fig. 1c and **Note 2**).

3. Gently tap the side of the dish and rinse the medium that is contained in the dish several times over the cells to gently remove the clusters of cells from the plate. Place this medium containing the iPSC clusters into a deep 10 cm² petri dish.

4. Perform this procedure two more times with 5 ml of fresh EIM to gently remove the clusters of cells from the plate and transfer medium containing iPSC clusters into the petri dish.

5. Add a final concentration of 5 μM of ROCK inhibitor (Y27632) to the petri dish containing EIM to ensure cell survival and efficient EB formation. Gently swirl dish to ensure adequate mixing of ROCK inhibitor.

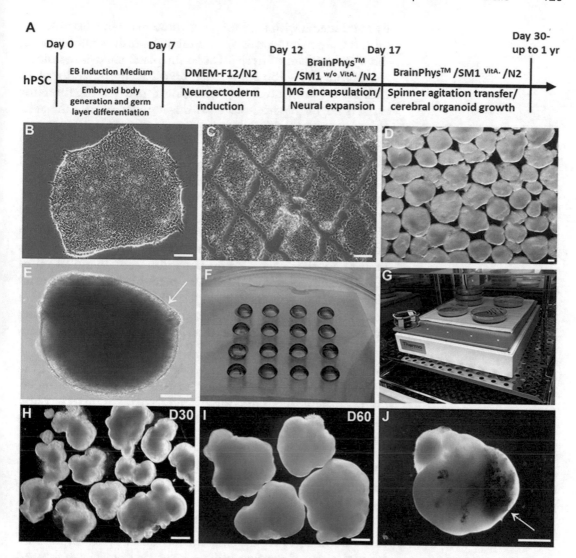

Fig. 1 Timeline and key developmental milestones for generating human cerebral organoids. (**a**) Schematic diagram overviewing the process to generate human cerebral organoids. (**b**) A hiPSC colony showing pluripotent characteristics such as compact growth and epithelial morphology. (**c**) Passaging of hiPSCs after using a StemPro® EZPassage™ tool showing uniform cutting of hiPSC colonies. (**d**) Day 2 EBs show aggregate formation and uniform size distribution. (**e**) A representative image of a day 10 EB differentiating towards neuroectoderm and displaying transparent neuroepithelium (arrow). (**f**) Matrigel spheres that are used to encapsulate neuroepithelial bodies. (**g**) Orbital shaker that is housed within a tissue culture incubator used to agitate encapsulated neuroepithelial bodies. (**h**) Day 30 and (**i**) day 60 cerebral organoids show growth up to 1–2 mm and 3–4 mm in diameter size, respectively. (**j**) A cerebral organoid containing retinal pigmented epithelium. Scale bars in panels **b**–**e** are equal to 100 μM; scale bars in panels **h**–**j** are equal 1 mm

6. Place petri dish within a humidified tissue culture incubator overnight that is set at 37 °C and 5% CO_2.

7. After 24 h incubation, EBs should be formed and should contain clear, smooth borders. Replace medium by pipetting up

EBs and media with a 25 ml pipette and place in a 50 ml conical tube. Remaining EBs within the petri dish is not an issue; simply add 10 ml of fresh EIM to the plate. Centrifuge pipetted EBs at 200 g for 2 min, aspirate off medium, and replace with 10 ml fresh medium containing 5 μM ROCK inhibitor using a 25 ml pipette. Scale up the number of 10 cm^2 petri dishes as appropriate ensuring no more than 100 EBs are cultured per dish.

8. Replace medium every other day for a total of 5 days, but only include ROCK inhibitor in the medium for the first 3 days (*see* Fig. 1d).

3.2 Neuroectoderm Induction of EBs

1. Culture EBs until they have grown to approximately 500 μm in diameter. Optimal EBs will have smooth and bright edges. Prepare another deep 10 cm^2 petri dish containing 15 ml of NIM. Remove optimal EBs from petri dish using a transfer pipet that has an opening of at least 1 mm in diameter into a petri dish containing NIM. Change NIM medium every 2 days by carefully pipetting and discarding half of the medium (~7.5 ml) and replacing with ~7.5 ml fresh medium.

2. Repeat the above process for 5–6 days and closely monitor the neuroectodermal differentiation of EBs. Optimal EBs undergoing efficient differentiation should display transparent outer surfaces reminiscent of neuroepithelium formation (*see* Fig. 1e and **Note 4**).

3.3 Encapsulation of Neuroepithelial Bodies Within Matrigel Spheres

1. Thaw as many tubes of Matrigel as needed on ice. One 500 μl aliquot will be sufficient to encapsulate approximately 15 neuroepithelial bodies.

2. While Matrigel is thawing on ice, work in a class 2 biological safety hood and prepare parafilm substrates by cutting a 4 in. by 4 in. piece of parafilm using scissors that have been sprayed with 70% ethanol to prevent contamination. Prepare as many parafilm pieces as necessary as each square will be sufficient to embed 16 neuroepithelial bodies.

3. Spray parafilm squares with 70% ethanol and let dry in hood.

4. By using the surface of a sterile and open 200 μl pipette box, perform indentations in the parafilm squares with the end of a 200 μl pipette tip that has been cut slightly. Indent parafilm with a grid of 4 rows and 4 columns of dimples to generate a total of 16 dimples. Place one piece of indented parafilm grid in the bottom of a petri dish and ensure parafilm lies flat by applying pressure to all of the edges.

5. By using a cut 1 ml pipette with an opening at least 1.5 mm in diameter, gently pipette one neuroepithelial body into the tip and then hold pipettor upright until body settles to the bottom of the pipette by gravitational force.

6. Place neuroepithelial body gently within one of the dimples of the parafilm grid. Repeat process until all 16 dimples contain neuroepithelial bodies.

7. Carefully remove residual medium from each dimple using a 200 µl pipettor without disrupting neuroepithelial bodies.

8. Aliquot 40 µl of Matrigel into each of the dimples containing neuroepithelial bodies (*see* Fig. 1f).

9. With the aid of a phase contrast light microscope, carefully position each neuroepithelial body in the center of each Matrigel sphere using a 10 µl pipette tip.

10. Transfer the petri dish containing Matrigel-embedded neuroepithelial bodies into a 37 °C tissue culture incubator for 30 min to allow for complete Matrigel polymerization.

11. Remove petri dish from incubator and pipette 10 ml of COEM to the dish without disrupting the Matrigel spheres. By using sterile forceps, invert Parafilm grid and expose to medium. Use forceps to gently agitate Parafilm to remove the spheres into the medium. For spheres that resist dislodging from parafilm, use forceps to gently remove them.

12. Culture encapsulated neuroepithelial bodies in a 37 °C tissue culture incubator for 2 days to allow for expansion of neuroepithelium within the Matrigel matrix.

13. Replace COEM after 2 days by gently aspirating (~5 ml) medium from the top of the petri dish by tilting dish and allowing Matrigel spheres to drop to the bottom. Carefully pipette 5 ml of fresh COEM into dish without disrupting Matrigel spheres.

14. Continue culturing encapsulated neuroepithelial bodies for an additional 2 days until transferring to an orbital shaker.

3.4 Differentiation of Cerebral Organoids with Spinning Agitation

1. Replace COEM with COGDM (similar to COEM, but containing vitamin A) by gently aspirating all medium from the top of the petri dish by tilting dish and allowing Matrigel spheres to drop to the bottom. Carefully pipette 15 ml of fresh COGDM into dish without disrupting Matrigel spheres.

2. Transfer dish containing organoids to an orbital shaker and set shaker speed at 50 rpm for the remainder of their growth and differentiation (*see* Fig. 1g and **Note 6**).

3. Change COGDM every 4 days and periodically monitor morphology using a stereomicroscope. By day 30, cerebral organoids can range between 1 and 2 mm in diameter (*see* Fig. 1h) and by day 60, larger and more complex organoids can develop up to 4 mm in diameter (*see* Fig. 1i). In a subset of organoids, extensive retinal pigmented epithelium can develop as denoted in Fig. 1j.

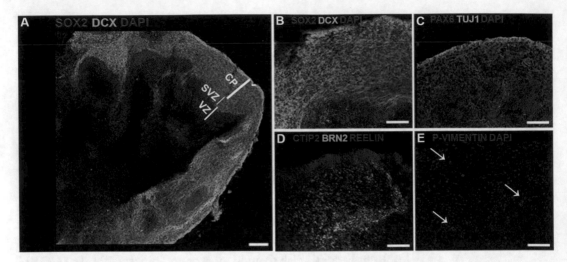

Fig. 2 Cerebral organoids display progenitor zone organization, cortical plate morphology, and produce oRGs at day 60. (**a**) Immunofluorescence staining of day 60 cerebral organoids show VZ and SVZ morphology as shown by expression of SOX2. DCX, a newborn neuronal marker, shows expression predominately on the outer surface of the organoid and delineates CP identity. (**b**) Immunofluorescence staining for SOX2 under high power magnification shows expression within the VZ and SVZ, and DCX expression is observed in the CP. (**c**) An additional progenitor cell marker, PAX6, stains forebrain progenitor zones and TUJ1 is expressed in neuronal processes. (**d**) Cortical layer 1 marker, REELIN, is expressed at the pre-plate and stains Cajal–Retzius neurons. CTIP2, cortical layer 5 marker, and BRN2, cortical layer 3 marker, is also expressed in day 60 organoids. (**e**) p-VIMENTIN staining is indicative of mitotically active radial glial cells (arrow). Scale bars in all panels are equal to 100 μM

3.5 Marker Characterization of Day 60 Cerebral Organoids

Multiple immunostaining approaches can be utilized to characterize the development of cerebral organoids at designated time points. By approximately 2 months of differentiation, key cortical developmental features can be observed such as the presence of progenitor zones, which are marked by SOX2 expression within the ventricular zone (VZ) and subventricular zone (SVZ) (*see* Fig. 2a, b). In addition, forebrain radial glial progenitor marker, PAX6, is observed in these zones as well (Fig. 2c). Cerebral organoids also exhibit cortical plate expression markers such as Doublecortin (DCX) and layer specific markers such as layer 1 marker, REELIN, layer 3 marker, BRN2, and layer 5 marker, CTIP2 (*see* Fig. 2d). Human-specific ORGs are also evident as shown by the expression of a phosphorylated form of VIMENTIN (*see* Fig. 2e).

4   Notes

1. The success of generating robust cerebral organoid cultures is critically dependent upon the quality of iPSC cultures. iPSC colonies should show hallmark signs of pluripotency such as clear colony borders, compact cell growth, and lack of differentiation. Common tests for pluripotency can include alkaline

phosphatase staining, and immunohistochemical staining for pluripotency markers such as OCT4 and NANOG. In addition, the TaqMan® hPSC Scorecard™ Assay tests for pluripotency and differentiation potential of all three germ lineages using real-time PCR analysis.

2. It is useful to use grid paper that is placed under the petri dish to ensure that the tool has consistently scored the entire plate in one direction. A permanent marker can be used to mark the top, bottom, and both sides of the dish to assist in orientation and proper cutting of iPSC colonies.

3. Typically by day 3, EBs will have grown to an adequate size that centrifugation is not necessary. Simply allow the EBs to settle to the bottom of the 50 ml conical tube by gravitational force. Carefully aspirate off medium leaving behind a few milliliters of medium, and be careful not to aspirate close to the EBs.

4. Timing is an important consideration in generating optimal cerebral organoids. Do not prolong differentiating EBs into neuroepithelium as this will affect the overall morphology and internal structure of organoids.

5. It is important to work as efficiently as possible during the encapsulation phase as the Matrigel will start to solidify within 10 min of being exposed to room temperature [11, 20].

6. It is critical to use an orbital shaker specifically designed to function in a tissue culture incubator that can withstand high humidity, temperature, and CO_2 levels.

Acknowledgments

This research was supported by Internal Research Funds by The Research Institute at Nationwide Children's Hospital.

References

1. Fietz SA, Huttner WB (2011) Cortical progenitor expansion, self-renewal and neurogenesis-a polarized perspective. Curr Opin Neurobiol 21(1):23–35. doi:10.1016/j.conb.2010.10.002

2. Gaspard N, Bouschet T, Hourez R, Dimidschstein J, Naeije G, van den Ameele J, Espuny-Camacho I, Herpoel A, Passante L, Schiffmann SN, Gaillard A, Vanderhaeghen P (2008) An intrinsic mechanism of corticogenesis from embryonic stem cells. Nature 455(7211):351–357. doi:10.1038/nature07287

3. Gotz M, Huttner WB (2005) The cell biology of neurogenesis. Nat Rev Mol Cell Biol 6(10):777–788. doi:10.1038/nrm1739

4. Manzini MC, Walsh CA (2011) What disorders of cortical development tell us about the cortex: one plus one does not always make two. Curr Opin Genet Dev 21(3):333–339. doi:10.1016/j.gde.2011.01.006

5. Chwalek K, Tang-Schomer MD, Omenetto FG, Kaplan DL (2015) In vitro bioengineered model of cortical brain tissue. Nat Protoc 10(9):1362–1373. doi:10.1038/nprot.2015.091

6. Schwartz MP, Hou Z, Propson NE, Zhang J, Engstrom CJ, Santos Costa V, Jiang P, Nguyen BK, Bolin JM, Daly W, Wang Y, Stewart R, Page CD, Murphy WL, Thomson JA (2015) Human pluripotent stem cell-derived neural

constructs for predicting neural toxicity. Proc Natl Acad Sci U S A 112(40):12516–12521. doi:10.1073/pnas.1516645112

7. Otani T, Marchetto MC, Gage FH, Simons BD, Livesey FJ (2016) 2D and 3D stem cell models of primate cortical development identify species-specific differences in progenitor behavior contributing to brain size. Cell Stem Cell 18(4):467–480. doi:10.1016/j.stem.2016.03.003

8. Qian X, Nguyen HN, Song MM, Hadiono C, Ogden SC, Hammack C, Yao B, Hamersky GR, Jacob F, Zhong C, Yoon KJ, Jeang W, Lin L, Li Y, Thakor J, Berg DA, Zhang C, Kang E, Chickering M, Nauen D, Ho CY, Wen Z, Christian KM, Shi PY, Maher BJ, Wu H, Jin P, Tang H, Song H, Ming GL (2016) Brain-region-specific organoids using mini-bioreactors for modeling ZIKV exposure. Cell 165(5):1238–1254. doi:10.1016/j.cell.2016.04.032

9. Hattori N (2014) Cerebral organoids model human brain development and microcephaly. Mov Disord: Off J Mov Disord Soc 29(2):185. doi:10.1002/mds.25740

10. Kadoshima T, Sakaguchi H, Nakano T, Soen M, Ando S, Eiraku M, Sasai Y (2013) Self-organization of axial polarity, inside-out layer pattern, and species-specific progenitor dynamics in human ES cell-derived neocortex. Proc Natl Acad Sci U S A 110(50):20284–20289. doi:10.1073/pnas.1315710110

11. Lancaster MA, Knoblich JA (2014) Generation of cerebral organoids from human pluripotent stem cells. Nat Protoc 9(10):2329–2340. doi:10.1038/nprot.2014.158

12. Lindborg BA, Brekke JH, Vegoe AL, Ulrich CB, Haider KT, Subramaniam S, Venhuizen SL, Eide CR, Orchard PJ, Chen W, Wang Q, Pelaez F, Scott CM, Kokkoli E, Keirstead SA, Dutton JR, Tolar J, O'Brien TD (2016) Rapid induction of cerebral organoids from human induced pluripotent stem cells using a chemically defined hydrogel and defined cell culture medium. Stem Cells Transl Med 5(7):970–979. doi:10.5966/sctm.2015-0305

13. Muzio L, Consalez GG (2013) Modeling human brain development with cerebral organoids. Stem Cell Res Ther 4(6):154. doi:10.1186/scrt384

14. Kirwan P, Turner-Bridger B, Peter M, Momoh A, Arambepola D, Robinson HP, Livesey FJ (2015) Development and function of human cerebral cortex neural networks from pluripotent stem cells in vitro. Development 142(18):3178–3187. doi:10.1242/dev.123851

15. Mariani J, Simonini MV, Palejev D, Tomasini L, Coppola G, Szekely AM, Horvath TL, Vaccarino FM (2012) Modeling human cortical development in vitro using induced pluripotent stem cells. Proc Natl Acad Sci U S A 109(31):12770–12775. doi:10.1073/pnas.1202944109

16. Pasca AM, Sloan SA, Clarke LE, Tian Y, Makinson CD, Huber N, Kim CH, Park JY, O'Rourke NA, Nguyen KD, Smith SJ, Huguenard JR, Geschwind DH, Barres BA, Pasca SP (2015) Functional cortical neurons and astrocytes from human pluripotent stem cells in 3D culture. Nat Methods 12(7):671–678. doi:10.1038/nmeth.3415

17. Shi Y, Kirwan P, Livesey FJ (2012) Directed differentiation of human pluripotent stem cells to cerebral cortex neurons and neural networks. Nat Protoc 7(10):1836–1846. doi:10.1038/nprot.2012.116

18. Hester ME, Murtha MJ, Song S, Rao M, Miranda CJ, Meyer K, Tian J, Boulting G, Schaffer DV, Zhu MX, Pfaff SL, Gage FH, Kaspar BK (2011) Rapid and efficient generation of functional motor neurons from human pluripotent stem cells using gene delivered transcription factor codes. Mol Ther: J Am Soc Gene Ther 19(10):1905–1912. doi:10.1038/mt.2011.135

19. Hester ME, Song S, Miranda CJ, Eagle A, Schwartz PH, Kaspar BK (2009) Two factor reprogramming of human neural stem cells into pluripotency. PLoS One 4(9):e7044. doi:10.1371/journal.pone.0007044

20. Lancaster MA, Renner M, Martin CA, Wenzel D, Bicknell LS, Hurles ME, Homfray T, Penninger JM, Jackson AP, Knoblich JA (2013) Cerebral organoids model human brain development and microcephaly. Nature 501(7467):373–379. doi:10.1038/nature12517

21. Shamir ER, Ewald AJ (2014) Three-dimensional organotypic culture: experimental models of mammalian biology and disease. Nat Rev Mol Cell Biol 15(10):647–664. doi:10.1038/nrm3873

22. Ranga A, Gjorevski N, Lutolf MP (2014) Drug discovery through stem cell-based organoid models. Adv Drug Deliv Rev 69-70:19–28. doi:10.1016/j.addr.2014.02.006

Chapter 9

Development of Mouse Cell-Based In Vitro Blood-Brain Barrier Models

Malgorzata Burek, Ellaine Salvador, and Carola Y. Förster

Abstract

Development of in vitro blood-brain barrier (BBB) models in the past four decades has provided the basis for intensive neurovascular research. However, handling and cultivating brain microvascular endothelial cells (BMEC) outside their natural environment is still a challenge. One of the strategies to get a stable BMEC culture is to generate an immortalized in vitro BBB model. This chapter describes how to isolate and immortalize mouse BMEC. A step-by-step protocol starting from the homogenization of brain tissue, digestion steps, seeding, and immortalization of the cells is given. Usually, it takes about 5 weeks to obtain a homogenous, immortalized BMEC line which can be used up to 50 passages.

Key words Blood-brain barrier, In vitro cell culture models, Immortalized brain microvascular endothelial cells

1 Introduction

The BBB consisting of BMEC, astrocyte end-feet, pericytes, and the basal membrane is responsible for the homeostasis and protection of the brain parenchyma. Studies of the BBB function at the cellular level are mostly based on in vitro BBB models. A considerable number of different in vitro BBB models have been established for research in different laboratories to date. Usually, the cells are obtained from bovine, porcine, rat, or mouse brain tissue [1]. Human tissue samples are limited due to ethical standards and availability of healthy material [2, 3]. A promising strategy in generating human in vitro BBB models was developed by using human stem cell lines. To date, three human stem cell-based in vitro BBB models have been described in the literature [4–6]. Primary BMEC isolations from rat or mouse tissue have to be done repeatedly because of small amounts of material. To overcome the problem of time-consuming isolations, different methods of cell line immortalization and establishment have been developed. Immortalizations using the simian virus 40 (SV40) large T antigen or adenovirus

Amit K. Srivastava et al. (eds.), *Stem Cell Technologies in Neuroscience*, Neuromethods, vol. 126, DOI 10.1007/978-1-4939-7024-7_9, © Springer Science+Business Media LLC 2017

early region 1A (E1A) have been successfully used to generate endothelial cell lines [7–9]. Moreover, the overexpression of human reverse transcriptase (hTERT) is being used for generation of human cell lines from different tissues. One prominent example for hTERT immortalized BMEC cell line is the well-characterized hCMEC/D3 [3, 10]. This cell line was generated from female epilepsy patient material. In addition to hTERT, hCMEC/D3 was immortalized with SV40 large T antigen.

The protocol presented here takes an advantage of polyoma middle T antigen (PymT) from murine polyomavirus. Murine polyomavirus induces endotheliomas in vivo and in vitro [11, 12]. The BMEC, once isolated and immortalized, should be evaluated for purity. Staining the cells with markers for astrocytes, pericytes, and neurons (e.g., glial fibrillary acidic protein, platelet-derived growth factor receptor beta, and NeuN, respectively) will exclude contamination of EC culture with these cells. The cells need to be evaluated for the expression of endothelial and BBB markers (e.g., CD31, von Willebrand factor, occludin, claudin-5), transendothelial electrical resistance (TEER), and permeability values.

Two BMEC lines, the brain microvascular endothelial cell lines termed cEND (from cerebral cortex) and cerebEND (from cerebellar cortex), were isolated according to this procedure in the Förster Laboratory [13–15]. Other cell lines generated by the same procedure are the mouse lines b.END3 and b.END5 [16, 17]. Both cEND and cerebEND have spindle-shaped morphology, values for TEER varying from 300 to 800 Ω cm^2 and express occludin and claudin-5 at the tight junctions (TJ). CEND and cerebEND respond to glucocorticoid [13, 18–21] and estrogen treatment [22, 23] as well as to pro-inflammatory mediators, such as TNF alpha [14, 19, 20]. Moreover, effects of treatment with multiple sclerosis patient sera [24, 25] and hypoxia [26–28] were studied on these cells. The presence of Abc transporters Abcb1, Abcg2, and Abcc4 has been demonstrated in cerebEND on mRNA and protein level [28]. Oxygen/glucose deprivation (OGD) and co-culture with C6 astrocytoma led to changes in the Abc transporter activity [28]. Moreover, cultivation of cEND under OGD conditions decreased TEER and lowered the TJ protein expression [26]. The expression of glucose transporter, Glut-1, as well as expression of Na(+)-D-glucose co-transporter, Sglt1, was also detected on mRNA and protein level and appeared to be elevated due to OGD conditions [14, 27, 28]. The cEND and cerebEND cell lines can be considered as a good tool for studying the structure and function of the BBB, cellular responses of ECs to different stimuli, or interaction of the EC with lymphocytes or cancer cells.

2   Materials

Bovine serum albumin (BSA), purity >98%.

cEND growth medium: 450 ml DMEM, 10% FCS, 10 ml L-glutamine, 2% MEM-Kit, 2% NEAA, 10 ml natrium pyruvate, 50 U/ml penicillin/streptomycin (or commercially available Endothelial Cell Culture Medium, e.g., from Cell Biologics Inc.).

Collagen IV.

Collagenase/dispase.

Fetal calf serum (FCS).

L-Glutamine.

MEM Vitamin.

Na pyruvate.

Neomycin (G418).

Nonessential amino acids (NEA).

Penicillin/streptomycin.

Phosphate-buffered saline (PBS).

Polybrene (hexadimethrine bromide).

Puromycin.

Trypsin/EDTA or Accutase.

3   Methods

3.1 Isolation of Brain Microvessels

1. For each preparation use five to ten neonatal mice of either sex (3–5 days old).

2. Remove the brain immediately after killing a mouse, and remove the meninges and capillary fragments through rolling the brain on a sterile blotting paper. Collect the forebrains by removing the brain stem, cerebella, and thalami with forceps. Transfer the brain into a Petri dish containing the following solution: 15 mM HEPES (pH 7.4), 153 mM NaCl, 5.6 mM KCl, 2.3 mM $CaCl_2$x $2H_2O$, 2.6 mM $MgCl_2$x $6H_2O$, and 1% (w/v) BSA (hereafter referred to as buffer A) (*see* **Note 1**).

3. Unless otherwise indicated, all isolation procedures should be performed at room temperature (RT) (22–24 °C) under the laminar flow hood. Cut the brains in buffer A using a sterile scalpel. Pipette the tissue fragments up and down using a 10 ml pipette until no clumps appear. Transfer the suspension into a 50 ml Falcon tube and centrifuge at 250 g for 5 min at RT.

4. Discard the supernatant.

5. Dissolve the pellet in 4.5 ml buffer A with 1.5 ml of 0.75% (w/v) collagenase/dispase (Roche), and incubate for 45 min at 37 °C in a water bath (occasionally shaking) (*see* **Note 2**).

6. Meanwhile, prepare the 12-well plates by coating four wells with collagen IV (0.1 mg/ml dissolved in 50 mM acetic acid). Allow to adhere for 1 h.

7. Stop the digestion by addition of 15 ml ice-cold buffer A. Resuspend the pellet thoroughly.

8. Centrifuge the suspension at 250 × g for 10 min at RT.

9. Discard the supernatant.

10. To remove myelin, add 10 ml 25% (w/v) BSA (Sigma, purity >98%), and centrifuge at 1000 × g for 20 min at 4 °C (*see* **Note 3**).

11. Carefully discard the supernatant, dissolve the pellet in 15 ml buffer A, and transfer to a new Falcon tube.

12. Centrifuge the suspension at 250 × g for 5 min at RT. The resulting pellet contains the endothelial cells.

13. Wash the collagen IV- coated 12-well plate twice with PBS.

14. Resuspend the pellet in 4 ml of growth medium (DMEM containing 10% FCS, 50 U/ml penicillin/streptomycin, 1% L-glutamine), and plate the cell suspension into two wells, 2 ml to each well. Incubate for 45 min at 37 °C in a cell culture incubator (*see* **Note 4**).

15. Transfer the 2 ml medium containing non-adhered cells from **step 14** into two fresh wells. Fill up the wells with adhered cells from **step 14** with 2 ml fresh medium.

16. *Optional*: add the puromycin to a final concentration of 4 μg/ml, and incubate the cells for 24 h (*see* **Note 5**).

17. Change the medium the following day.

3.2 Immortalizing the Brain Microvascular Endothelial Cells

1. Cultivate the GP + E-86 Neo (GPENeo) [29, 30] mouse fibroblasts, secreting a replication-deficient virus with polyoma middle T oncogene in DMEM medium containing 10% heat-inactivated FCS, 50 U/ml penicillin/streptomycin, and 2 mg/ml G418 in gelatin-coated flasks (*see* **Note 6**) (Figs. 1 and 2).

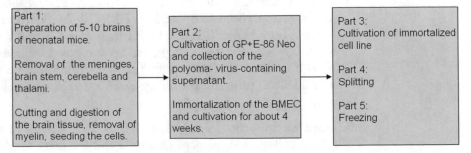

Fig. 1 Schematic representation of the described protocol. Brain microvascular endothelial cells (BMEC) are isolated from the cortex of neonatal mice brains. We describe in detail the cutting and digestion of brains (*Part 1*), immortalization of BMEC (*Part 2*), and cultivation, splitting, and freezing (*Parts 3–5*) of the immortalized cell line

2. To use the virus-containing medium for immortalization, cultivate the GPENeo cells in G418-free medium for 24 h.

3. Remove 10 ml of the GPENeo supernatant, and add Polybrene (hexadimethrine bromide) (Sigma) to a final concentration of 8 µg/ml, and sterilize through 0.45 µm filter to remove the cellular fragments.

4. Remove the growth medium from the EC cultures prepared in Part 1, and add 2 ml of GPENeo supernatant/Polybrene mixture into the wells. Repeat it the next day using a fresh GPENeo supernatant/Polybrene mixture (*see* **Note 7**).

5. Remove the GPENeo supernatant/Polybrene mixture the following day, wash the cells twice with PBS, and maintain the cells in the growth medium changing it every 3 days. Upon confluence, split the cells 1:2 on collagen IV-coated plates.

6. Typically, stable cerebral endothelial (cEND) cell lines should be obtained 4–5 weeks later.

3.3 Cultivating the Brain Microvascular Endothelial Cells

1. Thaw the cells from cryo-aliquot in water bath, and transfer in 15 ml Falcon tube with 10 ml pre-warmed medium.

2. Centrifuge the suspension at $250 \times g$ for 5 min at RT.

3. Remove medium, and transfer the pellet in the collagen IV-coated T25 cm^2 cell culture flask with pre-warmed growth medium (cell density for plating should be at least 1×10^4 cells/ml).

4. Change medium the next day, and after the cells reached confluence (usually after 5 days), split the cells.

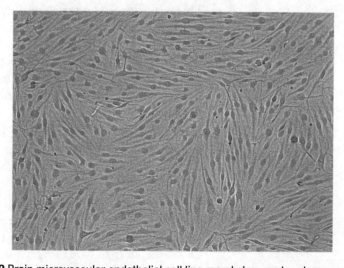

Fig. 2 Brain microvascular endothelial cell line morphology under phase contrast microscope. The cells show typical spindle-shaped morphology when cultured on collagen IV-coated plates

3.4 Splitting of cEND Cells (Note: Splitting Should Be Done Only Once a Week, Avoid Splitting Higher than 1:4.)

1. Remove medium and wash the cells with PBS.
2. Add 3 ml of warm trypsin-EDTA solution (T25 cm² flask), incubate at 37 °C, and wait until cell layer is dispersed (usually within 5–15 min) (*see* **Note 8**).
3. Add 5 ml of growth medium, pipette up and down, and transfer to a new collagen IV-coated flask.

3.5 Freezing of cEND Cells

1. Obtain the cell suspension as described in **step 3.4**.
2. Centrifuge the suspension at $250 \times g$ for 5 min at RT.
3. Resuspend the cell pellet (obtained from one T25 cm² flask) in 6 ml freezing medium (95% growth medium, 5% DMSO).
4. Divide the cell suspension into four 1.5 ml cryo-aliquots (*see* **Note 9**).
5. Store the cryo-aliquots under liquid nitrogen vapor temperature.

4 Notes

1. Meninges can be identified as thin, transparent membranes at the surface of the brain tissue. They can be carefully removed with sterile forceps or through rolling of the brain on a sterile blotting paper, e.g., Whatman paper.
2. Collagenase/dispase solution is prepared in buffer A.
3. 25% BSA should be dissolved in 1× PBS and filtered through 0.2 μm filter. We dissolve 25 g BSA in 100 ml PBS, let it stand overnight in the fridge, sterile-filter, and store it as frozen aliquots.
4. This step allows removal of rapidly adhering cells. The medium from step 1.14 contains non-adhered cells. These cells will be transferred into two fresh wells. This fraction contains the EC because they do not adhere within the 45 min incubation time of **step 1.14**. The cells which adhere fast (e.g., fibroblasts, astrocytes) will be grown in parallel and used for comparisons of morphology.
5. Puromycin is added for 24 h. Since only the brain ECs can metabolize puromycin, the other cell types will be dying during the puromycin treatment. This helps to get the pure EC culture [31].
6. GPENeo produces the replication-deficient virus which can specifically infect neonatal murine EC. Polyoma middle T oncogene (PymT) transfection causes growth advantage of ECs over non-ECs leading to a homogenous monolayer of cells with endothelial morphology 4–6 weeks of culture. The expression of PymT leads to immortalization of the EC cells and generation of endothelioma cell line. Comparisons of

endotheliomas and primary EC has shown that PymT-mediated immortalization leads to increased expression of genes responsible for the immune response due to viral origin of PymT. However, the expression of genes connected with the maintenance of barrier integrity and barrier properties was not influenced [32]. The viral vector produced by GPENeo contains the neomycin resistance gene. The immortalized cells will integrate this gene into their genome and will be therefore also neomycin resistant. This is to be considered for further experiments where the antibiotic selection is needed, e.g., by stable transfections. Moreover, the GPENeo fibroblasts should also be treated with neomycin prior to collection of viral particle-containing medium to be used for the immortalization of EC. Normally, treatment of GPENeo with 2 mg/ml neomycin for 2 weeks should be enough to maximize the amount of GPENeo cells producing the viral particles. GPENeo (GP + E-86 Neo) packaging cell line is commercially available at ATCC. GPENeo cell line producing PymT is available as shared published material [13, 33].

7. Polybrene is used to make pores in the cell wall to facilitate the infection with the viral particles.

8. The cells shouldn't be exposed to trypsin for too long, as it can destroy the cells. Sometimes already 2 min of trypsinization time is enough. Another possibility is to use Accutase for detachment of sensitive cells.

9. One cryo-aliquot can be seeded on T25 cm² flask.

Acknowledgments

This work was supported by the Deutsche Forschungsgemeinschaft (DFG, German Research Foundation) under grant number FO 315/4 and SFB 688.

References

1. Wilhelm I, Fazakas C, Krizbai IA (2011) In vitro models of the blood-brain barrier. Acta Neurobiol Exp (Wars) 71(1):113–128

2. Forster C, Burek M, Romero IA, Weksler B, Couraud PO, Drenckhahn D (2008) Differential effects of hydrocortisone and TNFalpha on tight junction proteins in an in vitro model of the human blood-brain barrier. J Physiol 586(7):1937–1949. doi:10.1113/jphysiol.2007.146852. jphysiol.2007.146852 [pii]

3. Weksler BB, Subileau EA, Perriere N, Charneau P, Holloway K, Leveque M, Tricoire-Leignel H, Nicotra A, Bourdoulous S, Turowski P, Male DK, Roux F, Greenwood J, Romero IA, Couraud PO (2005) Blood-brain barrier-specific properties of a human adult brain endothelial cell line. FASEB J 19(13):1872–1874. doi:10.1096/fj.04-3458fje. 04-3458fje [pii]

4. Lippmann ES, Azarin SM, Kay JE, Nessler RA, Wilson HK, Al-Ahmad A, Palecek SP, Shusta EV (2012) Derivation of blood-brain barrier endothelial cells from human pluripotent stem cells. Nat Biotechnol 30(8):783–791. doi:10.1038/nbt.2247. nbt.2247 [pii]

5. Boyer-Di Ponio J, El-Ayoubi F, Glacial F, Ganeshamoorthy K, Driancourt C, Godet M, Perriere N, Guillevic O, Couraud PO, Uzan G (2014) Instruction of circulating endothelial progenitors in vitro towards specialized blood-brain barrier and arterial phenotypes. PLoS One 9(1):e84179. doi:10.1371/journal.pone.0084179. PONE-D-13-16377 [pii]

6. Cecchelli R, Aday S, Sevin E, Almeida C, Culot M, Dehouck L, Coisne C, Engelhardt B, Dehouck MP, Ferreira L (2014) A stable and reproducible human blood-brain barrier model derived from hematopoietic stem cells. PLoS One 9(6):e99733. doi:10.1371/journal.pone.0099733. PONE-D-14-06808 [pii]

7. O'Connell KA, Edidin M (1990) A mouse lymphoid endothelial cell line immortalized by simian virus 40 binds lymphocytes and retains functional characteristics of normal endothelial cells. J Immunol 144(2):521–525

8. Roux F, Durieu-Trautmann O, Chaverot N, Claire M, Mailly P, Bourre JM, Strosberg AD, Couraud PO (1994) Regulation of gamma-glutamyl transpeptidase and alkaline phosphatase activities in immortalized rat brain microvessel endothelial cells. J Cell Physiol 159(1):101–113. doi:10.1002/jcp.1041590114

9. Ong SH, Dilworth S, Hauck-Schmalenberger I, Pawson T, Kiefer F (2001) ShcA and Grb2 mediate polyoma middle T antigen-induced endothelial transformation and Gab1 tyrosine phosphorylation. EMBO J 20(22):6327–6336. doi:10.1093/emboj/20.22.6327

10. Weksler B, Romero IA, Couraud PO (2013) The hCMEC/D3 cell line as a model of the human blood brain barrier. Fluids Barriers CNS 10(1):16. doi:10.1186/2045-8118-10-16. 2045-8118-10-16 [pii]

11. Kiefer F, Anhauser I, Soriano P, Aguzzi A, Courtneidge SA, Wagner EF (1994) Endothelial cell transformation by polyomavirus middle T antigen in mice lacking Src-related kinases. Curr Biol 4(2):100–109. doi:S0960-9822(94)00025-4 [pii]

12. Kiefer F, Courtneidge SA, Wagner EF (1994) Oncogenic properties of the middle T antigens of polyomaviruses. Adv Cancer Res 64:125–157

13. Forster C, Silwedel C, Golenhofen N, Burek M, Kietz S, Mankertz J, Drenckhahn D (2005) Occludin as direct target for glucocorticoid-induced improvement of blood-brain barrier properties in a murine in vitro system. J Physiol 565(Pt 2):475–486. doi:10.1113/jphysiol.2005.084038. jphysiol.2005.084038 [pii]

14. Silwedel C, Forster C (2006) Differential susceptibility of cerebral and cerebellar murine brain microvascular endothelial cells to loss of barrier properties in response to inflammatory stimuli. J Neuroimmunol 179(1–2):37–45. doi:10.1016/j.jneuroim.2006.06.019. S0165-5728(06)00254-2 [pii]

15. Burek M, Salvador E, Forster CY (2012) Generation of an immortalized murine brain microvascular endothelial cell line as an in vitro blood brain barrier model. J Vis Exp 66:e4022. doi:10.3791/40224022 [pii]

16. Brown RC, Morris AP, O'Neil RG (2007) Tight junction protein expression and barrier properties of immortalized mouse brain microvessel endothelial cells. Brain Res 1130(1):17–30. doi:10.1016/j.brainres.2006.10.083. S0006-8993(06)03166-0 [pii]

17. Yang T, Roder KE, Abbruscato TJ (2007) Evaluation of bEnd5 cell line as an in vitro model for the blood-brain barrier under normal and hypoxic/aglycemic conditions. J Pharm Sci 96(12):3196–3213. doi:10.1002/jps.21002

18. Forster C, Waschke J, Burek M, Leers J, Drenckhahn D (2006) Glucocorticoid effects on mouse microvascular endothelial barrier permeability are brain specific. J Physiol 573(Pt 2):413–425. doi:10.1113/jphysiol.2006.106385. jphysiol.2006.106385 [pii]

19. Burek M, Forster CY (2009) Cloning and characterization of the murine claudin-5 promoter. Mol Cell Endocrinol 298(1–2):19–24. doi:10.1016/j.mce.2008.09.041. S0303-7207(08)00445-0 [pii]

20. Forster C, Kahles T, Kietz S, Drenckhahn D (2007) Dexamethasone induces the expression of metalloproteinase inhibitor TIMP-1 in the murine cerebral vascular endothelial cell line cEND. J Physiol 580(Pt.3):937–949. doi:10.1113/jphysiol.2007.129007. jphysiol.2007.129007 [pii]

21. Blecharz KG, Drenckhahn D, Forster CY (2008) Glucocorticoids increase VE-cadherin expression and cause cytoskeletal rearrangements in murine brain endothelial cEND cells. J Cereb Blood Flow Metab 28(6):1139–1149. doi:10.1038/jcbfm.2008.2. jcbfm20082 [pii]

22. Burek M, Arias-Loza PA, Roewer N, Forster CY (2010) Claudin-5 as a novel estrogen target in vascular endothelium. Arterioscler Thromb Vasc Biol 30(2):298–304. doi:10.1161/ATVBAHA.109.197582. ATVBAHA.109.197582 [pii]

23. Burek M, Steinberg K, Forster CY (2014) Mechanisms of transcriptional activation of the mouse claudin-5 promoter by estrogen receptor alpha and beta. Mol Cell Endocrinol 392(1–2):144–151. doi:10.1016/j.mce.2014.05.003S0303-7207(14)00138-5. [pii]

24. Blecharz KG, Haghikia A, Stasiolek M, Kruse N, Drenckhahn D, Gold R, Roewer N, Chan A, Forster CY (2010) Glucocorticoid effects on endothelial barrier function in the murine

brain endothelial cell line cEND incubated with sera from patients with multiple sclerosis. Mult Scler 16(3):293–302. doi:10.1177/135 245850935818916/3/293. [pii]

25. Burek M, Haghikia A, Gold R, Roewer N, Chan A, Förster CY (2014) Differential cytokine release from brain microvascular endothelial cells treated with dexamethasone and multiple sclerosis patient sera. J Steroids Horm Sci 5:128–10.4172/2157-7536. 1000128

26. Kleinschnitz C, Blecharz K, Kahles T, Schwarz T, Kraft P, Gobel K, Meuth SG, Burek M, Thum T, Stoll G, Forster C (2011) Glucocorticoid insensitivity at the hypoxic blood-brain barrier can be reversed by inhibition of the proteasome. Stroke 42(4):1081–1089. doi:10.1161/STROKEAHA.110.5922 38STROKEAHA.110.592238. [pii]

27. Neuhaus W, Burek M, Djuzenova CS, Thal SC, Koepsell H, Roewer N, Forster CY (2012) Addition of NMDA-receptor antagonist MK801 during oxygen/glucose deprivation moderately attenuates the upregulation of glucose uptake after subsequent reoxygenation in brain endothelial cells. Neurosci Lett 506(1):44–49. doi:10.1016/j.neulet.2011.10. 045S0304-3940(11)01443-1. [pii]

28. Neuhaus W, Gaiser F, Mahringer A, Franz J, Riethmuller C, Forster C (2014) The pivotal role of astrocytes in an in vitro stroke model of the blood-brain barrier. Front Cell Neurosci 8:352. doi:10.3389/fncel.2014.00352

29. Golenhofen N, Ness W, Wawrousek EF, Drenckhahn D (2002) Expression and induction of the stress protein alpha-B-crystallin in vascular endothelial cells. Histochem Cell Biol 117(3):203–209. doi:10.1007/s00418-001-0378-7

30. Markowitz D, Goff S, Bank A (1988) A safe packaging line for gene transfer: separating viral genes on two different plasmids. J Virol 62(4):1120–1124

31. Perriere N, Demeuse P, Garcia E, Regina A, Debray M, Andreux JP, Couvreur P, Scherrmann JM, Temsamani J, Couraud PO, Deli MA, Roux F (2005) Puromycin-based purification of rat brain capillary endothelial cell cultures. Effect on the expression of blood-brain barrier-specific properties. J Neurochem 93(2):279–289. doi:10.1111/j.1471-4159.2004.03020.x. JNC3020 [pii]

32. Urich E, Lazic SE, Molnos J, Wells I, Freskgard PO (2012) Transcriptional profiling of human brain endothelial cells reveals key properties crucial for predictive in vitro blood-brain barrier models. PLoS One 7(5):e38149. doi:10.1371/journal.pone.0038149. PONE-D-12-03987 [pii]

33. Aumailley M, Timpl R, Risau W (1991) Differences in laminin fragment interactions of normal and transformed endothelial cells. Exp Cell Res 196(2):177–183

Chapter 10

Microfluidic Device for Studying Traumatic Brain Injury

Yiing Chiing Yap, Tracey C. Dickson, Anna E. King, Michael C. Breadmore, and Rosanne M. Guijt

Abstract

Throughout the world, traumatic brain injury (TBI), for example, as a result of motor vehicle accident, is a major cause of mortality and lifelong disability in children and young adults. Studies show that axonal pathology and degeneration can cause significant functional impairment and can precede, and sometimes cause, neuronal death in several neurological disorders including TBI, creating a compelling need to understand the mechanisms of axon degeneration. Microfluidic devices that allow manipulation of fluids in channels with typical dimensions of tens to hundreds of micrometers have emerged as a powerful platform for such studies due to their ability to isolate and direct the growth of axons. Here, we describe a new microfluidic platform that can be used to study TBI by applying very mild (0.5%) and mild (5%) stretch injury to individual cortical axons through the incorporation of microfluidic valve technology into a compartmented microfluidic-culturing device. This device is unique due to its ability to study the neuronal response to axonal stretch injury in a fluidically isolated microenvironment.

Key words Microfluidic, Stretch injury, Traumatic brain injury, Quake valve, Primary cell culture

1 Introduction

Traumatic brain injury (TBI) is the leading cause of mortality and morbidity in adults under 40 years of age, making TBI a significant public health problem [1]. Axonal injury is thought to be the most common pathology resulting from brain trauma [2]. It is proposed that the initial event in TBI is the stretching, compression, or breaking of white matter axons as a result of inertial forces commonly induced during traumatic incidents such as motor vehicle accidents, falls, violence, and sporting injuries [3–5]. This mechanical deformation progressively develops over a number of days triggering secondary events such as damage to the axonal cytoskeleton, particularly neurofilaments, causing regional compaction and/or impaired transport. As a result, accumulation of axon cargo induces regional swelling [6]. Furthermore, the primary changes in neurofilament structure

Amit K. Srivastava et al. (eds.), *Stem Cell Technologies in Neuroscience*, Neuromethods, vol. 126, DOI 10.1007/978-1-4939-7024-7_10, © Springer Science+Business Media LLC 2017

may be augmented by ionic shifts and secondary activation of proteases to degrade the axonal ultrastructure further [6]. Ultimately, axons may disconnect at the distal border of swollen regions, forming classic terminal bulbs, followed by Wallerian degeneration. Accordingly, the protracted response of the axon following TBI may provide a substantial "therapeutic window" for possible interventions to prevent, stop, or reverse the cellular changes ultimately leading to degeneration. Critical to the development of such interventions is a complete understanding of the cellular mechanisms that comprise the neuronal response to trauma, particularly those that underpin axonal degeneration.

Currently, there is no known long-term treatment that is available to fundamentally relieve or improve the outcome of TBI victims. The problem is compounded by the heterogeneity of brain injuries that can occur; the precise type, location, and extent of the primary injury; and the contribution of different pathological mechanisms and delayed secondary injury mechanisms. In addition, TBI is also a known risk factor for devastating neurodegenerative diseases including Alzheimer's, Parkinson's, and motor neuron disease [7]. Recent studies using cellular and animal models suggest axonal protection could be an important therapeutic strategy to improve symptoms in these diseases [8]. Thus, it is important to develop tools to elucidate the causes and effect of axonal injury and degeneration associated with TBI in order to provide valuable information on the mechanisms of degeneration which may ultimately lead to new treatments.

In this chapter, we describe a novel in vitro microfluidic model to simulate very mild (0.5%) and mild (5%) axonal stretch injury in a novel fluidically isolated compartmentalized culturing device [9]. This device consists of two independent polydimethylsiloxane (PDMS) structures separated by a 60 μm or 15 μm thick PDMS membrane (Fig. 1). Rat primary cortical neurons are grown in the upper PDMS microfluidic-culturing device (Xona Microfluidics), which has two microfluidic compartments of 100 μm height, 1.5 mm width, and 8 mm length interconnected with microgrooves of 10 μm width, 3 μm height, and 450 μm length. The small size of the microgrooves prevents migration of neuronal cell bodies between the compartments while allowing the axons to pass through [10]. The bottom structure contains a pneumatic channel (17 μm high, 90 μm wide, 40 mm long) and is irreversibly sealed with the PDMS membrane using a handheld corona discharge unit (Electro Technic Product Inc., USA). The pneumatic valve microfluidic device was replicated in PDMS by soft lithography and replica molding procedure from a patterned lithographic dry film master. In response to a controlled pressure pulse, the pneumatic channel inflates, and the PDMS membrane deflects upward, stretching the axons growing on top to varying degrees.

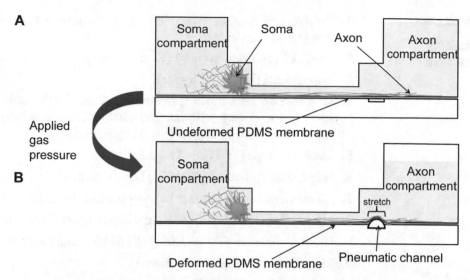

Fig. 1 Schematic drawing of the axonal stretch injury microfluidic device. (**a**) A thin PDMS membrane separates the pneumatic channel (*bottom*) from the overlying culturing chamber (*top*). (**b**) Application of gas pressure to the air channel (positioned at 200–300 μm from microgrooves) causes upward deflection of the thin PDMS, which stretches the overlying axon [9]

2 Materials

2.1 Equipment

1. Preparation of cell culture devices: Class II biosafety or laminar flow cabinet, sterile forceps, adhesive tape.

2. Primary cell dissection: Laminar flow cabinet, stereo dissecting microscope fitted with a cold-light source, water bath (37 °C), sterile dissection instruments (suggested array: 2× fine #5 forceps, 1× #3 forceps, hemocytometer).

3. Neuron culture: Humidified 5% CO_2 incubator, inverted cell culture phase-contrast microscope (Leica), fluorescence microscope (Leica).

4. Fabrication: Office laminator (Lamination System), programmable hot plate with 5 mm thick piece of polished aluminum plate (Torrey Pines Scientific), ultraviolet (UV) shark series high-flux LED array light source (Opto Technology), Laboratory Oven (Carbolite), 8″ portable precision spin coater (Cookson Electronics Equipment), handheld corona discharge unit (Electro Technic Product Inc.), vacuum-connected desiccator.

5. Characterization of device: Microscope with dual light-emitting diode (LED) light optical profiling in the vertical scanning interferometry (VSI) (Veeco Instruments Inc.).

6. Axonal stretch system: In-house built valve system containing a dynamic pressure transducer (Pneumadyne Inc.), USB-based controller for the 24 solenoid pneumatic valves (https://sites.google.com/site/rafaelsmicrofluidicspage/valve-controllers/usb-based-controller).

2.2 Materials

2.2.1 Materials for Microfabrication

1. Poly(methyl methacrylate) (PMMA) (75 mm × 50 mm × 1 mm) (RS component).

2. Ordyl 17 μm thick dry film (Elga Europe).

3. Isopropanol (EMD Millipore).

4. Transparency mask for pneumatic channel (90 μm wide, 40 mm long) (e.g., Kodak Polychrome image setting film Pagi-Set, 4400 dpi, Pagination Design Services).

5. BMR developer and rinse (Elga Europe).

6. Polydimethylsiloxane (PDMS) (Dow Corning).

7. 1.5 mm and 4 mm diameter biopsy punches (Huat Instrument).

8. 1H,1H,2H,2H–perfluorooctyltrichlorosilane (Fluorochem).

9. 100 mm diameter silicon wafer (SWI Semiconductor Wafer Inc.).

10. Razor blades (VWR International).

11. Weighing boat (VWR International).

12. 150 × 20 mm plastic dish (Sarstedt).

2.2.2 Preparation of Axonal Stretch Injury Device

1. Culturing microfluidic device with 450 μm wide microgrooves (Xona Microfluidics).

2. PDMS pneumatic channel device (make in-house).

3. 70% ethanol (EtOH).

4. Ultraviolet (UV) light.

5. Sterile 90 × 14 mm petri dish (VWR International).

2.2.3 Materials for Primary Cell Culture

1. Poly-L-lysine (PLL) (0.01% in PBS, Sigma-Aldrich). Store at 4 °C, and use within 1 month.

2. Neuron initial medium: Neurobasal medium (Invitrogen), heat-inactivated fetal bovine serum (10%, Invitrogen), B27 supplement (2%, Invitrogen), L-glutamine (0.5 mM, Gibco), L-glutamic acid (25 μM, Gibco), penicillin/streptomycin (1%, Invitrogen). Store at 4 °C, and use within 1 month.

3. Neuron subsequent medium: Neurobasal medium (Invitrogen), B27 supplement (2%, Invitrogen), L-glutamine (0.5 mM, Gibco), penicillin/streptomycin (1%, Invitrogen). Store at 4 °C, and use within 1 month.

4. Trypsin (0.25% stock, Invitrogen). Store aliquots at −20 °C, thaw before use, and do not freeze/thaw.

5. Trypan blue (Sigma-Aldrich).

6. Hanks' balanced salt solution (HBSS; Invitrogen).

7. 10 ml centrifuge tubes (Corning).

8. Sterile 90 × 14 mm petri dish (VWR International).

1. Triton X-100 (0.3% in PBS, Sigma-Aldrich). Store at 4 °C, and use within 1 month.

2. 4% paraformaldehyde (PFA) (Sigma-Aldrich). Danger, fixative reagent.

3. Primary antibodies: Tau (rabbit polyclonal, Dako), NFM (neurofilament M, rabbit polyclonal, Serotec), MAP2 (microtubule-associated protein 2, mouse monoclonal, Millipore).

4. Secondary antibodies: Alexa Fluor 488 (mouse IgG, Molecular Probes), Alexa Fluor 594 (rabbit IgG, Molecular Probes).

5. Fluorescent mounting medium (Immunotec).

3   Methods

3.1   Fabrication of PDMS Pneumatic Channel Device (Fig. 2)

See **Note 1**:

1. Remove the protective film of PMMA and wipe with lint-free tissue wetted with isopropanol, and then dry under a high-pressure air from a compressor.

2. Cut a piece of dry film photoresist to size and remove back protective film before placing the film on the PMMA sheet.

3. Laminate the PMMA substrate with the first layer of dry film at 40 °C at a speed of 1350 mm/min.

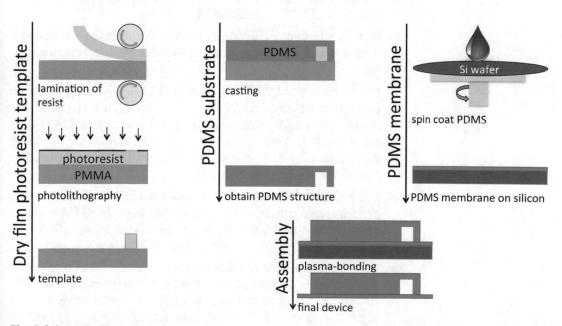

Fig. 2 Schematic illustration of the main steps of the fabrication process of the axonal stretch injury device replicating a template of photolithographically structured dry film photoresist on PMMA and plasma bonding this with a spin-coated PDMS membrane

4. Expose the entire sheet with UV shark series high-flux LED array for 2 min and then bake on hot plate for 2 min at 110 °C.

5. Cool the sheet to room temperature and remove the protective polyester layer.

6. Laminate a second layer of dry film resist onto the first layer of overexposed dry film.

7. Expose the entire sheet for 90 s using a transparency mask.

8. Bake the sheet on hot plate for 20 min at 110 °C, and then remove the protective film after cooling to room temperature.

9. Develop the pneumatic channel by submerging the sheet into BMR developer for 90 s and follow by rinsing in BMR rinse for 60 s.

10. Dry the developed template by blowing with compressed air.

11. Hard bake the template for 30 min at 110 °C on hot plate and then allow to cool to room temperature.

12. Place the dry film PMMA master template in a clean plastic petri dish (150 × 20 mm).

13. Prepare a mixture of PDMS elastomer and curing agent (10:1 ratio) in a weighing boat using a plastic fork, and then pour over the template (*see* **Note 2**).

14. Place inside the desiccator connected to a vacuum pump for 15 min to remove excessive bubbles.

15. Place the PDMS-poured PMMA dry film master in an oven for 60 min at 75 °C to cure the PDMS.

16. Detach the cured PDMS pneumatic device from the PMMA template and cut out around boundaries with a razor blade.

17. Silanize the silicon wafers with 1H,1H,2H,2H–perfluorooctyltrichlorosilane (PFOTC) prior to PDMS spinning to facilitate the release of PDMS from the silicon surface. Place the silicon wafer and tissue containing a few drops of PFOTC in the desiccator. Once a maximum vacuum level is attained, the desiccator valve is shut, and the silicon wafers are evaporation coated for 30 min (*see* **Note 3**).

18. Put a silanized silicon wafer onto vacuum chuck of the portable precision spin coater, and then spin coat with thin PDMS membrane at rate of 1500 rpm (~60 μm thickness) or 2500 rpm (~15 μm thickness) at spinning time of 30 s (*see* **Note 4**).

19. Treat both surface of PDMS membrane and PDMS pneumatic channel with air plasma for 15 s using handheld corona discharge unit, and then bond to each other and leave it overnight in oven at 75 °C to irreversibly seal these two pieces of PDMS and form final pneumatic device.

20. Punch the gas inlet of final PDMS pneumatic device with biopsy punch (1.5 mm diameter), and then irreversibly seal with another piece of punched PDMS (4 mm diameter hole) to form the inlet areas where the pneumatic channel layer can be accessible from the top via tubing (Fig. 3).

3.2 Computer-Controlled Opening and Closing of PDMS Pneumatic Valve by Pressure

1. Build an in-house valve system following the instructions using a dynamic pressure transducer, USB-based controller for the 24 solenoid pneumatic valves controlled by NI LabVIEW software.

2. Connect the pneumatic device inlets to the gas pressure source using Tygon tubing.

3. For opening and closing valves, pressure is controlled by a gas pressure line connected through pressure transducer to an array of solenoid valves.

4. The solenoid valves are connected through a USB interface by a computer and controlled using the National Instruments LabVIEW program that can be downloaded from https://sites.google.com/site/rafaelsmicrofluidicspage/valve-controllers/usb-based-controller.

5. Device operation and membrane deflection are visualized with a Veeco optical profiler system. This measurement provides the height of the deflection and enables calculation of the stretch injury using the half width of the channel using Pythagorean theorem. For example, a measured deflection of 5 μm of a 50 μm wide channel results in a $2 \times (\sqrt{(5^2 + 25^2)}) = 51$ μm long section or a stretch by $51–50 = 1$ μm. This corresponds to a $100 \times (1/50) = 2\%$ stretch (Fig. 4).

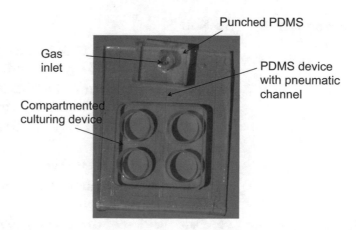

Fig. 3 Photograph of assembled device comprising pneumatic channel, gas inlet, and compartmented culturing device

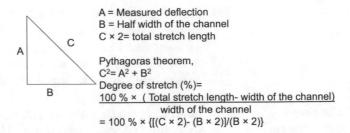

A = Measured deflection
B = Half width of the channel
C × 2= total stretch length

Pythagoras theorem,
$C^2 = A^2 + B^2$
Degree of stretch (%)=
$100\% \times \dfrac{(\text{Total stretch length- width of the channel})}{\text{width of the channel}}$
$= 100\% \times \{[(C \times 2) - (B \times 2)]/(B \times 2)\}$

Fig. 4 The calculation of the percentage of the stretch injury (%) that uses the half width of the pneumatic channel through Pythagorean theorem

3.3 Sterilization and Assembly of Axonal Stretch Injury Devices Prior to Culture

1. Clean surface particles of both culturing microfluidic devices and pneumatic device using adhesive tape.

2. Sterilize both devices for 30 min using UV light of biosafety hood, and then perform the subsequent steps under sterile condition.

3. Hydrophilize the surface of PDMS pneumatic device with handheld air plasma unit for 30 s, and then attach with compartmented culturing microfluidic device. Use blunt forceps to firmly press the culturing device onto the pneumatic device (*see* **Note 5**).

4. Place the final assembled device into a 90 mm diameter sterile plastic petri dish.

5. Immediately pipette 200 μL PLL into two top compartments of culturing microfluidic device (*see* **Note 6**).

6. Incubate at least 3 days at room temperature for adequate coating.

7. Remove PLL and fill the device with neuron initial plating medium. Allow equilibrating in cell culture incubator for a minimum of 24 h prior to addition of cells (*see* **Note 7**).

3.4 Primary Cortical Neuron Culture

See **Note 8**:

1. Sacrifice pregnant E17.5–18.5 rat (finding of a positive vaginal plug is defined E0.5) by CO_2 suffocation.

2. Spray lower abdomen with 70% EtOH and carefully remove embryos from the uterus using a pair of scissors.

3. Cut embryos (4–5) from amniotic sac and decapitate and place all heads on ice.

4. Dissect brain from the skull.

5. Dissect cortices from the brain into a small drop of HBSS and then remove the meninges.

6. Place the cleaned cortices into 10 ml sterile conical tubes containing 5 ml HBSS.

7. Repeat steps 4–6 with remaining heads (*see* **Note 9**).

8. Add trypsin to the final concentration 0.0125% (250 μL trypsin for 5 ml HBSS) to chemically dissociate the cells, and flick tube gently to mix.

9. Incubate with trypsin for 5 min at 37 °C water bath.

10. Remove trypsin solution from cells and wash with 2 ml neuron initial plating medium.

11. Resuspend cortical neurons with 500 μL of neuron initial plating medium, and then gently triturate cells 10–20 times with a 1 ml pipette tip until no visible cell clumps are present; however, avoid introducing air bubbles during trituration.

12. Perform cell count using a hemocytometer and trypan blue dye exclusion assay. If necessary, adjust volume of cell suspension to achieve 9×10^6 cells per ml.

13. Gently pipette 15 μL of cortical cell suspension to one reservoir of the soma side of the culturing microfluidic device, by pointing the tip of the pipette to the entrance of the cell chamber (*see* **Note 10**).

14. Return devices to the incubator for 5 min to allow cell attachment.

15. After incubation, fill medium reservoirs with warmed neuron initial plating medium and return cells to the incubator (*see* **Note 11**).

16. After 1 day in vitro (DIV), replace the medium with warmed neuron subsequent growth medium.

17. Main the culture media (with neuron subsequent medium) three times a week (on DIV 3, 5, 7) to prevent oxygen and nutrient depletion or waste accumulation.

3.5 Axonal Stretch Injury

1. Culture the neurons inside the axonal stretch injury device for DIV 7 to allow axons to grow through the microgrooves and then across the pneumatic channel (Fig. 5).

2. Connect the pneumatic channel inlet to gas pressure source (refer to Sect. 3.2) prior to axonal stretch injury.

3. Apply 35 psi gas pressure to a 60 μm thick PDMS membrane for 10 s to achieve 0.5% stretch and 20 psi gas pressure to a 15 μm thick PDMS membrane for 10 s to achieve 5% stretch (*see* **Note 12**).

3.6 Analysis and Characterization of the Response of Neuronal Culture to Axonal Stretch Injury

Live cell images are captured using an inverted phase-contrast cell culture microscope (Fig. 5), and immunocytochemistry images are visualized with a fluorescent microscope (Fig. 6).

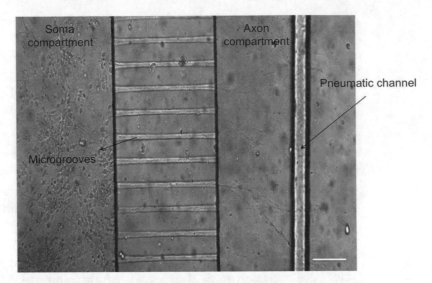

Fig. 5 Primary cortical neurons at DIV 7 in the PLL-coated culturing device showing adequate axonal extension prior to axonal injury. Scale bar = 150 μm [9]

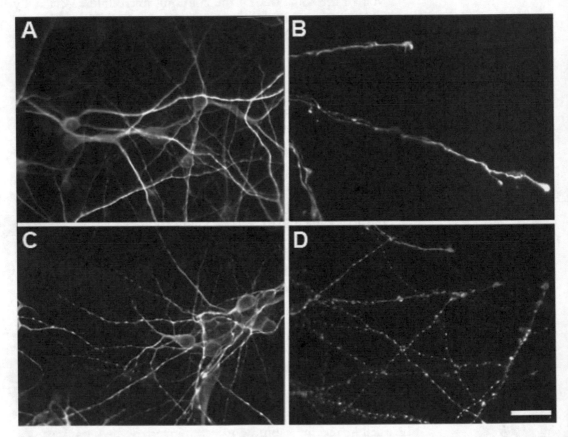

Fig. 6 Immunocytochemistry images of uninjured and injured neurons 24 h following 0.5% injury. Cell bodies and dendrites (**a**, MAP2 labeling), and axons (**b**, tau and NFM labeling), in the control chambers were smooth and uniformly labeled for cytoskeletal markers. At 24 h post-injury, cortical cultures exposed to axonal stretch injury showed dendritic blebbing and irregular MAP2 expression (**c**, MAP2 labeling). The injured axons (**d**, tau and NFM labeling) underwent characteristic beading and degeneration, showing punctate accumulation of tau and NFM within the swollen portions of the axon. Scale bar = 50 μm [9]

1.1.1 Fixation
and Immunocytochemistry

1. Aspirate culture medium from all reservoirs.

2. Fill one reservoir on each side with 4% PFA for 1 h in room temperature.

3. Incubate on orbital shaker during fixation.

4. Remove fixative and wash the cells with 0.01 M PBS.

5. Use forceps to carefully remove culturing microfluidic devices from pneumatic device.

6. Permeabilize the cells on the pneumatic device with 0.3% Triton X-100 in 0.01 M PBS for 15 min at room temperature.

7. Label the cells with neuron-specific primary antibodies such as anti-tau (1:5000), anti-neurofilament M (NFM, 1:1000), or microtubule-associated protein 2 (MAPs 2, 1:1000) for 1 h at room temperature and 4 °C for overnight, and then incubate with isotype and specific secondary antibodies (Alexa Fluor 488 or 594) for 2 h at room temperature (*see* **Note 13**).

8. Wash the cells in 0.01 M PBS three times (10 min each time) and then mount on glass slide with fluorescent mounting medium and then allow to air-dry in dark at room temperature.

9. Capture the fluorescent images using a fluorescent inverted microscope.

4   Notes

1. For the fabrication of the template, steps 1–12 may be replaced by alternative microfabrication processes available to the researchers, for example, a SU-8 template on glass or silicon using photolithography process.

2. Improper mixing of polymer base reagent and curing agent can result in unevenly cured polymer.

3. PFOTC is toxic, so all these steps should be done in chemical hood.

4. The PDMS needs to first spin coat at 500 rpm for 10 s to spread the solution thoroughly over the silicon wafer prior to the required spinning speed. Thickness of the final layer depends on the viscosity of the PDMS prepolymerization mixture and hence on the temperature in the laboratory.

5. Make sure devices are completely sealed to each other to avoid leakage. Besides, pneumatic channel is positioned at 200–300 µm from microgrooves with the aid of microscopy.

6. Minimize confusion between soma and axon side of devices by marking with permanent marker on underside of axonal stretch injury device. Minimize bubbles when adding PLL.

7. Can be left in incubator for up to 3 days; however, remember to make sure devices do not dry out.

8. All experiments involving animals were conducted according to protocols approved by the Animal Ethics Committee of the University of Tasmania.

9. Minimize cell loss by performing dissection within 90 min of obtaining tissue. Cell viability will be reduced if longer time is taken to obtain tissue.

10. Cells should be visibly attached to the chamber within the first few minutes of addition. Cells still readily flowing >10 min after addition indicate insufficient substrate coating.

11. Adding medium prior to cell attachment will flush cells from growth chamber.

12. The degree of pressure can be adjusted by using the pressure transducer. The LabVIEW software can be used to set the duration of pressure pulse.

13. Anti-tau or anti-NFM is used to visualize axon morphology because tau is an axon-specific microtubule-associated protein and NFM is a neurofilament protein which is highly expressed in some axons. MAP 2 is a dendrite-specific microtubule-associated protein and is used to visualize dendrites. However, the types of antibodies can be replaced depending on the end purpose, e.g., antibodies to synaptophysin can be used for the visualization of presynapses.

References

1. Sosin DM, Sniezek JE, Waxweller RJ (1995) Trends in death associated with traumatic brain injury, 1979 through 1992-sucess and failure. JAMA 273:1778–1780
2. Povlishock JT (1992) Traumatically induced axonal injury-pathogenesis and pathobiological implications. Brain Pathol 2:1–12
3. Adams JH, Doyle D, Graham DI et al (1984) Diffuse axonal injury in head injuries caused by fall. Lancet 2:1420–1422
4. Grady MS, McLaughlin MR, Christman CW et al (1993) The use of antibodies targeted against the neurofilament subunits for detection of diffuse injury in humans. J Neuropathol Exp Neurol 52:143–152
5. Meaney DF, Smith DH, Shreiber DI et al (1995) Biochemical analysis of experimental diffuse axonal injury. J Neurotrauma 12:689–694
6. Povlishock JT, Erb DE, Astruc J (1992) Axonal response to traumatic brain injury- reactive axonal change, differentiation and neuroplasticity. J Neurotrauma 9:S189–S200
7. Coleman MP, Perry VH (2002) Axon pathology in neurological disease: a neglected therapeutic target. Trends Neurosci 25:532–537
8. Raff MC, Whitmore AV, Finn JT (2002) Neuroscience - axonal self-destruction and neurodegeneration. Science 296:868–871
9. Yap YC, Dickson TC, King AE, Breadmore MC, Guijt RM (2014) Microfluidic culture platform for studying neuronal response to mild to very mild axonal stretch injury. Biomicrofluidics 8:044110
10. Taylor AM et al (2005) A microfluidic culture platform for CNS axonal injury, regeneration and transport. Nat Methods 2:599–605

Chapter 11

Multimodal Neural Stem Cell Research Protocols for Experimental Spinal Cord Injuries

Yang D. Teng, Evan Y. Snyder, Xiang Zeng, Liquan Wu, and Inbo Han

Abstract

Human neural stem cells (hNSCs) have been the focus in basic science and translational research as well as in investigative clinical applications. Therefore, the capability to perform reliable derivation, effective expansion, and long-term maintenance of primordial and uncommitted NSCs in vitro is essential for growing the capacities of stem cell biology and regenerative medicine. In this chapter, we systematically summarized a set of protocols and unique procedures that have been developed in the laboratories of Dr. Teng, Dr. Snyder, and their collaborators. These regimens have been, over years, productively used to derive, propagate, maintain, and differentiate operationally defined human somatic NSCs. We emphasize our established multimodal methodologies for charactering the *functional multipotency* of stem cells and their value in basic as well as translational scientific studies.

Key words Stem cell, Neural stem cell, Mesenchymal stromal stem cell, Polymer, Synthetic material, Biodegradation, Functional multipotency, Spinal cord injury, Stroke, Transplantation

1 Murine Neural Stem Cells (mNSCs)

Cells from the CNS with stemlike properties have been successfully isolated from the embryonic, neonatal, and adult murine CNS. They can be propagated in vitro by many effective and safe means that include both epigenetic and genetic strategies. The epigenetic approach includes mitogens such as epidermal growth factor (EGF) [1] or basic fibroblast growth factor (bFGF or FGF2) [2]. The genetic method consists of gene transfer with propagating genes such as v*myc*, *large T-antigen* (*T-Ag*), or *telomerase* [3–5]. Importantly, maintaining murine NSCs (mNSCs) in a proliferative state in culture does not appear to subvert their ability to respond to normal developmental cues in vivo following transplantation. Upon entering the in vivo environment, NSCs generally withdraw from the cell cycle, interact with host cells, and differentiate to express cellular markers for neuronal and glial lineages [6, 7]. Often, a small subpopulation will survive in long terms and remain

Amit K. Srivastava et al. (eds.), *Stem Cell Technologies in Neuroscience*, Neuromethods, vol. 126, DOI 10.1007/978-1-4939-7024-7_11, © Springer Science+Business Media LLC 2017

as quiescent and undifferentiated cells intermixed seamlessly among host cells [8]. These extremely plastic cells of both rodents and human intend to migrate and differentiate in temporal and regional manners in response to cue molecules, which emulates endogenous NSCs following implantation into germinal zones as well as acts like white blood cells to migrate into lesion/inflammatory areas based on chemotaxic mechanisms (e.g., SDF-1α and CXCR4, VEGF and VEGFR-1, etc.) [8–11]; in addition, they can express developmental cytokines including trophic or anti-inflammatory factors and engineered foreign genes that are homeostatic/therapeutic, shed exosomes, form gap junctions, and are capable of neural cell replacement (i.e., functional multipotency of NSCs) [6–8, 12]. For example, multipotent, clonal mNSC lines, generated from neonatal murine cerebellum, can participate in normal development along the rodent neuraxis, largely independent of the initial region from which they were isolated, attesting to their stemness. Donor mNSCs can integrate not only back into the developing cerebellum but throughout the entire neuraxis of the immature and adult rodent central and peripheral nervous system in a non-tumorigenic, cytoarchitecturally matching manner [13]. Differentiation fate appears to be determined by site-specific microenvironmental factors in addition to neural progenitor lineage; cells from the same clonal line differentiated into neurons or glia based on their site of engraftment. Transplanted NSCs that differentiated into ultrastructurally identifiable neurons appeared, for example, to participate in synaptogenesis, suggesting functional as well as anatomic integration [13, 14]. Similar phenomena have been reported for adult NSCs obtained from rodent hippocampus [15], spinal cord [16], ventricular-subventricular zone (V-SVZ) [17], and cortex [18, 19]. Therefore, NSCs, as modeled by these various mNSC lines, can be used as informative investigative probes and have the capability for repair of, or transport of genes into, the CNS. Furthermore, such cellular level plasticity provides a unique mechanism for a developmental strategy whereby multipotent progenitors migrate with commitment to cell phenotype occurring amid interaction with their microenvironment. While mouse and rat generally seem to show similar biological behaviors under most circumstances, the accumulated data suggest that mouse NSCs are mostly more practical than those of rat to propagate and maintain for prolonged periods for as-yet undetermined reasons. However, most technological approaches and biological concepts derived from mouse NSCs can be extrapolated to rat NSCs.

2   Maintenance and Propagation of Murine NSCs In Vitro

Note: all cell culture medium, trophic factors, cytokines, genetic molecules, and other experimental reagents can be ordered from American Type Culture Collection (Manassas, VA, USA), Atlanta

Biologicals (Flowery Branch, GA, USA), EMD Millipore-Life Science (Billerica, MA, USA), Innovative Cell Technologies, Inc. (San Diego, CA, USA), Life Technologies (Grand Island, NY, USA), Invitrogen (Carlsbad, CA, USA), Thermo Fisher Scientific Inc. (Hudson, NH, USA), Sigma-Aldrich Corporation (St. Louis, MO, USA), Promega (Madison, WI, USA), or other vendors with validated product reputation.

The most widely used genetically engineered mNSC line in our laboratories and that of our collaborators is C17.2 that was initially developed by Dr. E. Y. Snyder and colleagues [13]. C17.2 cells, a prototypical NSC clone, are originally derived from neonatal mouse cerebellum as previously described [13].

2.1 Methods and Materials

1. Derivation of NSC lines can be most readily accomplished by dissecting the primary structure of interest (e.g., cerebellum, olfactory bulb, cortex, hippocampus, spinal cord) from immature (fetal or newborn) mice and incubating the tissue in 0.5% trypsin in phosphate-buffered saline. If the tissue is abstracted from adult animals, dissociation will require collagenase and/or Accutase® as well as longer incubation times.

2. Tissue is triturated and washed twice before cultures as plated on uncoated tissue culture dishes in serum-containing medium. Serum at this stage both inactivates the trypsin and insures the health of the dissociated cells.

3. At this stage, cells can then be propagated in serum-free medium with mitogens–EGF (10–20 ng/ml) plus FGF2 (10–20 ng/ml) plus LIF (10 ng/ml)–or they can be transduced with a gene that enhances propagation by acting downstream of these mitogens. For the growth factor selection process of NSCs, please see section on human NSCs below.

4. If one elects to use the genetic approach to propagate NSC clones, then the primary culture described above is infected 24–48 h after plating, by incubation with the avian *myc* or viral *myc* (v*myc*) vector PK-VM-2. pneoMLV is the parent retrovirus vector in which v*myc* DNA was inserted. Cells are then cultured in DME + 10% FCS + 5% horse serum (HS) + 2 mM glutamine for 3–7 days, until cultures appear to have undergone at least two doublings.

5. Cultures are then trypsinized (or accutased) and seeded at 3–10% confluence in DME + 10% FCS + 0.3 mg/ml G418. Neo-resistant colonies are typically observed within 1–2 weeks.

6. Chosen colonies are then isolated by brief exposure to trypsin within plastic cloning cylinders after ~2 weeks. Colonies are then replated and expanded on uncoated 24-well CoStar™ plates. At confluence, these cultures are further passaged and expanded to first 35 mm and then 60 mm and ultimately

100 mm uncoated Corning® tissue culture dishes. Cells infected with control vectors without propagating genes did not survive beyond this passage in serum-containing medium.

7. Early passages of the expanded colonies were frozen in Nunc cryostat tubes at 10^6 cells/vial in DME + 30% FCS + 12% DMSO (this method of freezing is also effective for most other types of murine NSC lines).

2.2 Genetic Modification

Immortalized NSC lines such as C17.2 can then be transfected with additional genes, such as lacZ, the gene encoding *E. coli* ß-galactosidase (β-Gal).

1. This procedure is done by plating a recent 1:10 split of the cell line of interest onto 60 mm tissue culture plates. 24–28 h after plating, the cells are incubated with a lacZ-encoding retroviral vector (e.g., BAG) plus 8 μg/ml Polybrene for 1–3 h. Cells are then cultured in fresh feeding medium (DME + 10% FCS + 5% HS + 20 mM glutamate) for ~3 days until they will have undergone through at least two doublings.

2. The cultures are then trypsinized (or "accutased") and seeded at low density (50–5000 cells on a 100 mm tissue culture dish). If the newly inserted transgene contains a selection marker distinct from your first selection marker, e.g., non-neomycin-based selection marker such as hygromycin or puromycin, then one can select for the new infectants as described above.

3. After ~3 days, well-separated colonies are isolated by brief exposure to trypsin (or Accutase®) within plastic cloning cylinders, as described above. If the additional transgene does not have a new selection marker or also employs *neo*, identification of infectants is still possible. Select colonies as described and proceed to plate them in 24-well plates. At confluence, these cultures can be passaged to and expanded in increasingly larger surface area tissue culture dishes.

4. A representative dish from each clone can then be stained histochemically for your new gene of interest, e.g., for ß-Gal expression via Xgal histochemistry [7, 8, 13], directly in the culture dish. The percentage of Xgal + blue cells is then assessed microscopically. The clones with the highest percentage of blue cells are then used for future studies.

   (a) The ability to detect the lacZ gene product by Xgal histochemistry in vivo and/or by anti-β-Gal immunohistochemistry makes such lacZ-expressing cells effective for transplantation studies, especially when it is used under cross verification of immunocytochemistry deploying antibody specifically against ß-Gal [7, 8, 11].

Preferred murine cerebellar or other type of NSC lines were grown in DME + 10% FCS + 5% HS + 2 mM glutamine on tissue culture dishes coated with poly-L-lysine (PLL). They are either fed weekly with 1/2 conditioned medium from confluent cultures +1/2 fresh medium, or split (1:10) weekly into fresh medium. Lines should not routinely be split more dilutely than 1:10 although one may potentially have the ability to do so. We recommend that for important transplant studies, a given thaw of cells should not have been passaged for longer than 4–8 weeks. It is better to thaw out an earlier vial and use one of the two or three post-expansion vials for the planned in vivo study. Therefore, a stockpile of as many as possible of early freezes is advisable. Subclones can be abstracted by seeding 50–5000 cells on 10 cm dishes in the typical medium or in 1/3 conditioned medium from confluent cultures. Well-separated colonies can again be isolated using cloning cylinders. One should be aware that, for reasons that still remain not specified, changes in brands of tissue culture plastic, type of substrate coat, and type, concentration, and lot of serum can change the phenotypes displayed by NSC lines.

3   Preparation of Murine NSCs In Vitro for In Vivo Transplantation

For preparation of cells for transplantation in, for example, the injured spinal cord of rodents, here is a succinct description of procedure workflow.

3.1   Spinal Cord Injury Models

1. Young adult or adult Sprague-Dawley (or a different strain) rats (body weight, 225–250 g) are randomly assigned for receiving different pre- and/or post-injury treatments with group size carrying adequate power of statistics [20]. All in vivo protocols should be in accordance with the principal investigator's Institutional Animal Care and Use Committee (IACUC) guidelines and policies issued by federal and local regulatory agencies. For each SCI (spinal cord injury) site, midline contusion [20] or compression [21] or hemicontusion [22] or segmental hemisection [8] is created using a standardized injury device or size 11 surgical scalpel, under IACUC-approved sufficient anesthesia. Hemostasis is normally achieved by using Gelfoam (Pfizer, New York, NY). An independent observer who is blinded for the experimental group design should confirm the adequacy and consistency of the lesion severity or length and breadth of the lesion. Only at that time is the surgeon informed of the treatment (e.g., cell suspension, solution injection, implanting of NSC-seeded polymer construct, etc.) to be administered to the lesion. Following either the full or control treatment, the musculature was sutured, skin closed, and the rodent recovered in a cage with clean bedding materials on a heating pad till it fully awakes (usually about 4 h). Ringer's lactate solution (0.5–1.0 ml/mouse, 5.0–

10.0 ml/rat) should be given daily for 5–7 days post-operation. When injury affects spontaneous micturition reflex, bladders were gently evacuated twice daily until a so-called "reflex bladder" function is established [20].

3.2 Preparation and Quality Control of Donor NSCs

1. NSC preparation: a subconfluent (75–85%, not >90%) 10 cm culture dish of adherent mNSCs (or cells of the line of interest) is washed with PBS (1–3 times).

2. Cautions: if cells are allowed to become too confluent for more than 24–48 h prior to transplantation, then they begin to elaborate an extracellular matrix that causes them to become clumpy and renders poor engraftment and often autonomous clusters of cells. Also, the cells begin to exit the cell cycle and differentiate, which also predisposes to poor engraftment. The more immature and proliferative and well isolated the cells, the better the engraftment.

3. Recommendations: of course, it is reassuring to purposely allow sister plates to proceed (in parallel assessments in vitro) to confluence and differentiation to insure that the cells still do possess the ability to differentiate into mature neural phenotypes. It is better to use many dishes of 90% confluent cells than a fewer number of dishes with more cells that have become heavily confluent.

4. Cells that have been prepared appropriately for transplantation (we usually do a split 48 h before the intended grafting procedure) are then gently trypsinized (or accutased), smoothly but thoroughly triturated into well-dispersed single-cell suspensions and then resuspended in PBS (pH, 7.4) or in feeding medium plus ~0.05% w/v trypan blue to aid with localization of the injected material) to give a high cellular concentration (3–6×10^4 cells/μl) which is nevertheless free-flowing and will not obstruct the glass pipette (Φ, 70–100 μm) or needle used for implantation.

5. Although we employ finely drawn glass micropipettes using Flaming/Brown Micropipette Puller (Model: P-97; Sutter Instrument Co. Novato, CA, USA), one can also employ a Hamilton syringe, though we recommend connection of a glass micropipette, instead of a conventional metal needle, with it for cell injections. It is important to minimize cell settlement out of suspension (which can occur rather quickly if the suspension solution is not prepared correctly) and to ensure that there is no clump formation that is usually detected by sharply increased resistance to injection advancement. Hence, we maintain the cells on ice and gently triturate them regularly. One should be careful not to fall into the trap of implanting cell carrying vehicle from which the cells have settled out yet believing that one has implanted cells. Interpretation of data, as one might imagine, will be entirely erroneous.

3.3 Donor Cell Tracking

For NSCs engineered via retrovirus to express a transgene, e.g., for ß-Gal-transduced cell lines, the "helper virus-free" status needs to be demonstrated prior to transplantation. Otherwise, host-derived cells may be falsely interpreted as being donor derived. The test can be done by confirmation of the failure of supernatants from the cell lines to produce neomycin-resistant colonies or Xgal-positive colonies in naive cells (e.g., 3T3 fibroblasts). In addition, transgene+ cells should be systematically validated to exclude possibilities of cell fusion-mediated contamination before confirming their true identity.

While the most efficient method for detecting cells in vivo is via the presence of a transgene, the risk of downregulation of a transgene, for unspecified/unclear reasons and thus usually unpredictably, presents enough of a problem that multiple simultaneous methods of donor cell detection in vivo should also be put in place prior to an important transplantation study.

1. For example, FISH (fluorescence in situ hybridization) against the *lacZ* gene itself can be effective. Using NSCs derived from a male animal and implanting them in a female animal allows one to do FISH against the Y-chromosome.

2. Most easily, one can preincubate the NSCs in BrdU or EdU ~48 h (depending on cell doubling time) prior to transplantation and subsequently identify the cells in vivo through the use of an anti-BrdU antibody (which requires DNA denaturation by typically using HCl, heat, or digestion with DNase to expose the BrdU for immuno-detection).

3. Or a simple EdU detection assay (which is based on a click reaction via a copper-catalyzed reaction between an azide-containing detection reagent and an EdU possessing alkyne). The small size of azide allows the use of mild conditions to readily detecting EdU incorporated into the DNA. Either assay will reveal donor-derived cells via their immunopositive nucleus [7]. Most NSCs only undergo 0–2 cell divisions in vivo following transplantation, usually not enough to dramatically "dilute" the EdU or BrdU marker (e.g., it takes five symmetric divisions to get rid of ~97% BrDU in each cell).

4   Human NSCs (hNSCs)

Similar to their rodent counterparts, stable clones of NSCs have been isolated from the human fetal telencephalon [5, 7, 23–26]. These self-renewing clones in vitro give rise to all fundamental neural lineages. Following transplantation into germinal zones of the brain or the spinal cord of new born rodents, they participate in aspects of normal development, including migration along established migratory pathways to disseminated CNS regions, intended differentiation or differentiation into multiple developmentally and regionally appropriate cell types, and nondisruptive interspersion with host progenitors and their progeny. Prototype or genetically engineered ex vivo, human NSCs (hNSCs) are capable of expressing innate or foreign transgenes in vivo in these disseminated locations [5, 23, 24]. The soluble products secreted from and/or gap junctions formed by these hNSCs can cross-correct an inherited genetic metabolic defect in neurons and glia in vitro or in vivo, further supporting their gene therapy potential. Finally, human NSCs have potential to replace specific deficient neuronal populations [5, 7, 23, 24, 27].

For research and potential clinical applications, hNSCs have been studied and found feasible for cryopreservation. In addition, these cells can be propagated by both epigenetic and genetic means that are comparably safe and effective. The observations encourage investigations of NSC transplantation for a range of disorders. Thus we have tested various culture conditions and genes for those that optimally allow for the continuous, efficient expansion and passaging of prototype hNSCs [7, 11, 28]. For genetic engineering modifications, v*myc* (e.g., the p110 gag-myc fusion protein derived from the avian retroviral genome) seems to be one of the most effective genes (see the following sections for more details). In either case, we have also identified a strict requirement for the presence of mitogens (FGF2 and EGF) in the growth medium, in effect constituting a conditional perpetuality or immortalization (LIF has been effective for blunting potential senescence) [7, 26, 28, 29].

4.1 hNSC Protocols

1. In general, our monoclonal, nestin-positive, human neural stem cell lines perpetuated in this way divide every ~40 h and stop dividing upon mitogen removal, undergoing spontaneous morphological differentiation and upregulating markers of the three fundamental lineages in the CNS (neurons, astrocytes, and oligodendrocytes) [28, 29].

2. Therefore, hNSC lines (e.g., HFB2050) retain basic features of epigenetically expanded human neural stem cells [7, 11, 29, 30]. Clonal analysis confirmed the stability, multipotency, and self-renewability of the cell lines.

3. Finally, hNSC lines can be transfected and transduced using a variety of procedures and genes encoding proteins for marking purposes (e.g., lacZ) and of therapeutic interest (e.g., BDNF) [13, 23, 31].

The isolation, propagation, characterization, cloning, and transplantation of NSCs from the human CNS appeared to share a similar tack established by prior experience with the successful murine NSC clone C17.2 (propagated following transduction of a constitutively downregulated v*myc* [13, 23, 31] and with growth factor-expanded murine NSC clones [32].

5 Induced Conditional Self-Renewing Human NSCs

For the current field of induced pluripotency or potency and fate reprogramming, an interesting question is that to what extent a cell should be reprogramed back to a more primitive state for translational purposes. The standard scenario requires reprogramming a terminally committed cell back to pluripotence before instructing it toward a particular specialized phenotype. The process is in general demanding and increases risks of developing neoplasia and undesired cell types. Since precursor/progenitor cells derived for therapeutic goals typically lack only one critical attribute – the capacity for sustained (and/or controllable) self-renewal. We deemed that the capacity could be induced in a regulatable manner such that cells proliferate only in vitro (or, for most cases, in short terms following transplantation) and differentiate in vivo without common dangers related to promoting pluripotence or specifying lineage identity.

As proof of concept, we theorized and tested the efficiency, safety, engraftability, and therapeutic utility of "induced conditional self-renewing progenitor cells" derived from the human CNS. In our study, the self-renewal capability was efficiently induced within neural progenitors solely by introducing v*myc* tightly regulated by a tetracycline (Tet) on gene expression system (details of methods in Kim et al.) [33]. Tet in the culture medium activated *myc* transcription and translation, allowing efficient expansion of homogeneous, clonal, karyotypically normal human CNS precursors ex vivo. After transplantation in vivo, when Tet was not administered, *myc* expression was absent, which impeded self-renewal and its tumorigenic potential. Cell proliferation ceased, and differentiation into electrophysiologically active neurons and other CNS cell types in vivo was facilitated upon transplantation into rats.

Besides showing the efficacy in both developmental rats and adult rats after ischemia injury (e.g., functional improvement and without neoplasia, donor cell overgrowth, emergence of non-neural cell types, phenotypic or genomic instability, etc.), this approach also mitigated the need for immunosuppression. We postulate that the strategy of "induced conditional self-renewal (ICSR)" might be applied to progenitors of other lineages. It may ultimately prove to be a safe, effective, efficient, and practical method for optimizing investigative or therapeutic potential gained from the ability to genetically reprogram stem or progenitor cells [33].

6 Isolation, Selection, Maintenance, and Propagation of Human NSCs In Vitro

1. A suspension of primary dissociated neural cells (5×10^5 cells/ml), prepared from the telencephalon (particularly the periventricular region) of an early second trimester human fetus (e.g.,

13 weeks or 15 weeks) [23, 25, 34], is plated on uncoated tissue culture dishes (Corning) first in serum-containing medium as described about for mouse for ~24–48 h and then in the following "growth medium": Dulbecco's Modified Eagles Medium (DMEM) + F12 medium (1:1) supplemented with N2 medium (Gibco) to which FGF2 (10–20 μg/ml) + heparin (8 μg/ml) and EGF (10–20 μg/ml) and LIF (10 ng/ml) are added.

2. Cultures are then put through the following "growth factor" selection process: cells are transferred to FGF2-containing serum-free medium alone for 2–3 weeks; then they are cultured in EGF-containing serum-free medium alone for another 2–3 weeks; consequently they are returned to FGF2-containing serum-free medium alone for 2–3 weeks.

3. Finally, they are maintained in serum-free medium containing FGF2 plus LIF (EGF is optional). At each stage of selection, large numbers of cells would die or fail to survive passaging. What would be left following the above selection process are passageable, immature, proliferative cells that are responsive to both FGF2 and EGF (i.e., co-expression of both receptors), qualities essential, in our view, for operationally defining an NSC.

4. Medium is changed every 5–7 days. Cell aggregates are dissociated in trypsin-EDTA (0.05%)/or Accutase® when ≥10–20 cell diameters in cluster or sphere sizes and replated in growth medium at 5×10^5 cells/ml.

5. To verify differentiation potential, some dissociated hNSCs were plated on poly-L-lysine (PLL)-coated slides (Nunc) in DMEM (devoid of growth factors) + 10% fetal bovine serum (FBS; one may add 2 mM L-glutamine and 1000 U/ml penicillin/1000 μg/ml streptomycin) and processed weekly for immunocytochemistry (ICC). In most cases, differentiation occurred spontaneously or under enhanced promotion of selected trophic factors [29].

6. For astrocytic maturation, clones can be co-cultured with primary dissociated embryonic CD-1 mouse brain [35]. As a new approach, neural cell (neuron, in particular) differentiation can be induced by culturing NSCs in nanofibers fabricated in unique patterns [36].

7. Polyclonal populations of the hNSCs are then separated into single clonal lines either by serial dilution alone (i.e., one cell per well–the process often has poor yields for prototypic hNSCs) or by first infecting the cells with a retrovirus (allowing there to be an engineered mitotic gene and/or molecular marker of clonality–i.e., the proviral integration site) and then performing serial dilution. The transgenes transduced via retrovirus are either (non-transforming) propagation enhancement genes, such as vmyc or human telomerase reverse transcriptase (hTERT) and/or purely reporter genes such as lacZ [5, 23].

8. For retrovirus-mediated gene transfer, two xenotropic, replication-incompetent retroviral vectors have been used to infect hNSCs. A vector encoding *lacZ* is similar to BAG [13] except for bearing a PG13 xenotropic envelope. An amphotropic vector encoding v*myc* was generated using the ecotropic vector described for generating murine NSC clone C17.2 [35] to infect the GP + envAM12 amphotropic packaging line [37]. No helper virus needs to be produced. Infection of FGF2-maintained human neural cells with either vector (titer, 4×10^5 CFUs) can follow similar, previously detailed procedures [13, 37].

9. For cloning of hNSCs, cells are dissociated as above, diluted to 1 cell/15 μl, and plated at 15 μl/well of a Terasaki or 96-well dish. Wells with single cells were noted immediately. Single-cell clones can be expanded and maintained in FGF2-containing growth medium. Single cells grow best when conditioned medium from dense hNSC cultures is included as at least 20–50% of the growth medium.

10. Monoclonality is confirmed by identifying in all progeny a single and identical genomic insertion site on Southern analysis for, as an example, either the *lacZ*- or the v*myc*-encoding provirus as previously detailed [35]. The v*myc* probe is generated by nick translation labeling with 32P dCTP; a probe to the *neo*-sequence of the *lacZ*-encoding vector is generated by PCR utilizing 32P dCTP.

11. Lastly, cryopreservation of hNSCs is done by resuspending post-trypsinized/accutased human cells in a freezing solution composed of 10% DMSO-, 50% FBS-, and 40% FGF2-containing growth medium. Afterward the temperature of cells is brought down slowly first to 4 °C for an hour and then to −80 °C for 24 h and then to −140 °C [38].

7 Preparation of Human NSCs In Vitro for In Vivo Transplantation

1. Prepare in vivo model properly as per the SCI modeling system described above.

2. Cultured cells grow as a combination of adherent cells and floating clusters within T25 flasks. All cells must be collected and be well dispersed into a suspension of individual cells in order for them to engraft efficiently.

3. To accomplish this, all medium and cells (including those adherent which are mechanically dislodged) are transferred into a 15 ml centrifuge tube and centrifuged for 3 min at 1000 rpm. Following removal of the supernatant, 0.7 ml of trypsin/EDTA (0.05% trypsin, 0.53 mM EDTA, or Accutase®,

typically 2.5–5 ml for a T25 flask depending upon confluency and density of the cell culture) is added to the centrifuge tube, and the cells are again triturated briefly before a 5-min incubation at 37 °C to facilitate dissociation of cells from each other.

Note: Further gentle triturating is required to break up pellets and reach a true single-cell suspension status (though trypsin can also be added to the original flask in order to retrieve cells that may have still been adherent to the flask, it must be used with caution as over-digestion could trigger irreversible cell clunking). Trypsinization is then terminated by adding 0.7 ml trypsin inhibitor (0.25 mg/ml in PBS) into the tube (or the flask) and triturating the mixture thoroughly. After a 3-min centrifuge at 1000 rpm (radius of rotor: 100 mm; $112 \times g$), and removal of supernatant, cells are washed 1–3 times by resuspending them in 10 ml PBS.

4. Procedures for additional labeling of cells with trackers such as DiI or Hoechst can be done at this stage according to protocols suggested by manufacturers. Then cells can be resuspended with small volume of PBS (or with a sufficient amount of trypan blue, e.g., ~0.05% w/v, to permit viability assay or localization of the injected suspension).

5. An ideal injection concentration of cells (i.e., $5–10 \times 10^4$ cells/ μl) can be achieved by cell counting using a hemocytometer for final volume adjustments. The remaining transplantation procedures are similar to those employed for murine NSCs.

8 Multimodal Applications of NSC-Based Investigative and Translational Approaches

As discussed in the previous sections, the transplantation of NSCs into areas of injury can be useful for cell replacement and/or for delivery of therapeutic genes. Some of the most impressive examples of this, in platform technology establishing studies, were observed in rat models of traumatic SCI and mouse models of hypoxic-ischemic (HI) cerebral injury [6, 8]. Although NSCs appear to have the potential to repopulate severely injured spinal cord or HI-injured brain, their ability to survive, reconstitute neural tissue, and reform neural connections is often limited by the vast amount of parenchyma loss and the consequent secondary injury cellular and molecular milieu. Since the epicenter of the primary lesion changes rapidly into a necrotic syrinx (the so-called cystic cavity or lesion cavity), even the most vital NSC may need an organization template that partially serves as extracellular matrix (ECM) skeletons to support survival and guide restructuring. In addition, large volumes of cells will not survive if located greater than a few 100 μm from the nearest capillary [6]. Hence, *we first hypothesized that three-dimensional highly porous (or purposely "structurally patterned") "scaffolds" composed of biodegradable (natural or synthetic) copolymers (or polymers) such as poly(lactic-co-glycolic acid) (or polyglycolic acid, PGA) coated with poly-L-lysine or other molecules that are also well tolerated by live biological systems, if pre-seeded with NSCs (or other types of cells) for cotransplantation into the infarction cavity or space in solid organs, might facilitate donor cell survival, migration, differentiation, and reformation of structural and functional circuits* [6, 8]. Since PLGA or PGA is a synthetic biodegradable polymer with Food and Drug Administration (FDA) approval for a variety of clinical applications, this approach has since become a primary platform technology for devising research and clinical applications of cells, especially stem cells. Highly hydrophilic, PLGA or PGA loses its mechanical strength rapidly over 2–4 weeks in the body; the scaffold can initially provide a matrix to guide cellular organization and growth, allowing diffusion of oxygen/nutrients to the transplanted cells, become vascularized, and then disappear, obviating concerns over long-term biocompatibility.

To test this hypothesis for spinal cord repair, a multicomponent, biodegradable, synthetic PLGA scaffold of specified architecture and seeded with NSCs was designed to support and structure neural repair, including possibilities of neural regeneration, direct cell replacement, impediment of glial scar formation, stabilization of blood-spinal cord barrier, and mitigation of secondary injury events [8]. The implantation of the scaffold seeded with NSCs in an adult rat segmental hemisection model of spinal cord injury (T9–T10; initial length, 4 mm) led to robust long-term improvement in function relative to the lesion control group. At 70 days post injury, the scaffold with cell group exhibited coordinated weight-bearing stepping as compared with movement of two to three hind limb joints in the lesion control group. Importantly, transplantation of scaffold alone also showed significantly benefit for locomotion improvement, with the treated rats demonstrating body weight-bearing stepping in the hind limbs. In contrast with conventional data, tract tracing revealed no corticospinal tract axons passing through the injury epicenter to the caudal side of the cord; there were, however, very few BDA (biotinylated dextran amine)-labeled CST fibers spotted along the interface between the injured side and the contralateral intact spinal cord, suggesting that they were likely spared by the NSC-seeded polymer or polymer-only treatment. Histological and immunocytochemical analysis including antibodies against donor-specific or general NSC markers (e.g., nestin), neurofilament (H, M, and L), GFAP, and GAP-43 indicated that functional recovery was the product of a significant mitigation of secondary tissue loss resulting in an increase in preserved tissue and reconstitution of host neural tissue of the damaged spinal cord. Most importantly, we observed no donor-derived mature neuronal replacement in the lesion epicenter, suggesting an overall suppressive environment for neurogenesis in the adult mammalian spinal cord; the few long surviving donor cells (~1–3%) remained to be progenitors showing nestin-positive immunoreactivity. *Therefore, we for the first time concluded that NSCs, in addition to their lineage differentiation capacity, could produce neurotrophic and anti-inflammatory factors/cytokines in responses to lesion niche demands to promote tissue and functional repair through rebuilding homeostasis* [8]. *This evidence, together with the finding that NSCs can nurture functioning neural networks via forming gap junctions with their surrounding cells, enabled us to engender a novel concept of stem cell biology – the functional multipotency of stem cells* [12, 38]. Our work showed that the broadened biology spectrum of stem cells including the essential capability of terminal phenotypic differentiation can be enhanced by polymer-mediated multimodal applications. For example, when NSC-PGA polymer combination was transplanted into the cerebral infarction cavity via a glass micropipette 4–7 days following induction of an experimental HI insult, the NSCs were observed to completely impregnate the PGA matrix, and the NSC-PGA unit refilled the infarction cavity. Part of the reconstituted tissue became vascularized by the host. The polymer-scaffolded NSCs displayed robust engraftment, foreign gene expression, and differentiation into neurons and glia within the region of HI injury, which, in contrast with the spinal cord, showed a neurogenesis permissive environment in the adult mammalian brain. Neuronal tracing assessment with DiI and BDA suggested that the long-distance neuronal circuitry between donor-derived and host neurons in both cerebral hemispheres might have been reformed through the corpus callosum in some cases [6].

These groundbreaking findings suggest that multimodal applications of stem cells (e.g., using drug-embedding polymer to promote cell survival/guide cell fate or stage "biological reactors") may facilitate even further the repair potential, function, and differentiation of donor and host neurons and augment the ingrowth/outgrowth of such cells to help promote reformation of structural/functional spinal cord and brain neural circuits. Together the studies done by our teams and others suggest that organ or iPSC-derived NSCs can play a fundamental role in neural repair via mechanistically oriented strategies of neurobiological investigation [8, 11, 29], tissue engineering [36], controlled drug release [28], conditional reprogramming [33], cell replacement [6, 7], and gene-directed enzyme prodrug therapy (GDEPT) [39]. Indeed, the data further reinforce the idea that, for CNS and PNS repair, NSCs may serve not only as an individual investigative or therapeutic vehicle but also as an anchor that holds concomitant approaches together: molecular and cell interaction, gene therapy, drug delivery, biomaterial, tissue engineering, niche modification, and cell replacement.

Acknowledgments

The work at Teng Laboratories has been supported by Project ALS, NIH, VA, DoD, BSF, the Gordon Project to Treat and Cure Clinical Paralysis, and CASIS-NASA.

Author contributions: YDT wrote and finalized the manuscript. XZ, IH, and LW actively participated in stem cell culture, cryoprotection, implantation preparation, and protocol refinements. The authors hold no COI for the paper.

References

1. Deleyrolle LP, Reynolds BA (2009) Isolation, expansion, and differentiation of adult mammalian neural stem and progenitor cells using the neurosphere assay. Methods Mol Biol 549:91–101

2. Qian X, Davis AA, Goderie SK, Temple S (1997) FGF2 concentration regulated the generation of neurons and glia from multipotent cortical stem cells. Neuron 18:81–93

3. Cepko CL (1989) Immortalization of neural cells via retrovirus-mediated oncogene transduction. Annu Rev Neurosci 12:47–65

4. Dottori M, Tay C, Hughes SM (2011) Neural development in human embryonic stem cells-applications of lentiviral vectors. J Cell Biochem 112:1955–1962

5. Roy NS, Chandler-Militello D, Lu G, Wang S, Goldman SA (2007) Retrovirally mediated telomerase immortalization of neural progenitor cells. Nat Protoc 2:2815–2825

6. Park KI, Teng YD, Snyder EY (2002) The injured brain interacts reciprocally with neural stem cells supported by scaffolds to reconstitute lost tissue. Nat Biotechnol 20:1111–1117

7. Redmond DE Jr, Bjugstad KB, Teng YD, Ourednik V, Ourednik J, Wakeman DR, Parsons XH, Gonzalez R, Blanchard BC, Kim SU, Gu Z, Lipton SA, Markakis EA, Roth RH, Elsworth JD, Sladek JR Jr, Sidman RL, Snyder EY (2007) Behavioral improvement in a primate Parkinson's model is associated with multiple homeostatic effects of human neural stem cells. Proc Natl Acad Sci U S A 104:12175–12180

8. Teng YD, Lavik EB, Qu X, Park KI, Ourednik J, Zurakowski D, Langer R, Snyder EY (2002) Functional recovery following traumatic spinal cord injury mediated by a unique polymer scaffold seeded with neural stem cells. Proc Natl Acad Sci U S A 99:3024–3029

9. Schmidt NO, Przylecki W, Yang W, Ziu M, Teng YD, Kim SU, Black PM, Aboody KS, Carroll RS (2005) Brain tumor tropism of transplanted human neural stem cells is induced by vascular endothelial growth factor. Neoplasia 7:623–629

10. Imitola J, Raddassi K, Park KI, Mueller FJ, Nieto M, Teng YD, Frenkel D, Li J, Sidman RL, Walsh CA, Snyder EY, Khoury SJ (2006) Directed migration of neural stem cells to sites of CNS injury by the stromal cell-derived factor 1alpha/CXC chemokine receptor 4 pathway. Proc Natl Acad Sci U S A 101:18117–18122

11. Teng YD, Benn SC, Kalkanis SN, Shefner JM, Onario RC, Cheng B, Lachyankar MB, Marconi M, Li J, Yu D, Han I, Maragakis NJ, Lládo J, Erkmen K, Redmond DE Jr, Sidman RL, Przedborski S, Rothstein JD, Brown RH Jr, Snyder EY (2012) Multimodal actions of neural stem cells in a mouse model of ALS: a meta-analysis. Sci Transl Med 4(165):165ra164

12. Teng YD, Yu D, Ropper AE, Li J, Kabatas S, Wakeman DR, Wang J, Sullivan MP, Redmond DE Jr, Langer R, Snyder EY, Sidman RL (2011) Functional multipotency of stem cells: a conceptual review of neurotrophic factor-based evidence and its role in translational research. Curr Neuropharmacol 9:574–585

13. Snyder EY, Deitcher DL, Walsh C, Arnold-Aldea S, Hartwieg EA, Cepko CL (1992) Multipotent neural cell lines can engraft and participate in development of mouse cerebellum. Cell 68:33–51

14. Taylor RM, Lee JP, Palacino JJ, Bower KA, Li J, Vanier MT, Wenger DA, Sidman RL, Snyder EY (2006) Intrinsic resistance of neural stem cells to toxic metabolites may make them well suited for cell non-autonomous disorders: evidence from a mouse model of Krabbe leukodystrophy. J Neurochem 97:1585–1599

15. Kempermann G, Song H, Gage FH (2015) Neurogenesis in the adult hippocampus. Cold Spring Harb Perspect Biol 7:a018812

16. Shihabuddin LS, Ray J, Gage FH (1997) FGF-2 is sufficient to isolate progenitors found in the adult mammalian spinal cord. Exp Neurol 148:577–586

17. Lim DA, Alvarez-Buylla A (2014) Adult neural stem cells stake their ground. Trends Neurosci 37:563–571

18. Florio M, Huttner WB (2014) Neural progenitors, neurogenesis and the evolution of the neocortex. Development 141:2182–2194

19. De Juan RC, Borrell V (2015) Coevolution of radial glial cells and the cerebral cortex. Glia 63:1303–1319

20. Teng YD, Choi H, Onario RC, Zhu S, Desilets FC, Lan S, Woodard EJ, Snyder EY, Eichler ME, Friedlander RM (2004) Minocycline inhibits contusion-triggered mitochondrial cytochrome c release and mitigates functional deficits after spinal cord injury. Proc Natl Acad Sci U S A 101:3071–3016

21. Ropper AE, Zeng X, Anderson JE, Yu D, Han I, Haragopal H, Teng YD (2015) An efficient device to experimentally model compression injury of mammalian spinal cord. Exp Neurol 271:515–523

22. Choi H, Liao WL, Newton KM, Onario RC, King AM, Desilets FC, Woodard EJ, Eichler ME, Frontera WR, Sabharwal S, Teng YD (2005) Respiratory abnormalities resulting from midcervical spinal cord injury and their reversal by serotonin 1A agonists in conscious rats. J Neurosci 25:4550–4559

23. Flax JD, Aurora S, Yang C, Simonin C, Wills AM, Billinghurst LL, Jendoubi M, Sidman RL, Wolfe JH, Kim SU, Snyder EY (1998) Engraftable human neural stem cells respond to developmental cues, replace neurons, and express foreign genes. Nat Biotechnol 16:1033–1039

24. Eriksson PS, Perfilieva E, Bjork-Eriksson T, Alborn AM, Nordborg C, Peterson DA, Gage FH (1998) Neurogenesis in the adult human hippocampus. Nat Med 4:1313–1317

25. Vescovi AL, Parati EA, Gritti A, Poulin P, Ferrario M, Wanke E, Frolichsthal-Schoeller P, Cova L, Arcellana-Panlilio M, Colombo A, Galli R (1999) Isolation and cloning of multipotential stem cells from the embryonic human CNS and establishment of transplantable human neural stem cell lines by epigenetic stimulation. Exp Neurol 156:71–83

26. Roy NS, Wang S, Jiang L, Kang J, Benraiss A, Harrison-Restelli C, Fraser RA, Couldwell WT, Kawaguchi A, Okano H, Nedergaard M, Goldman SA (2000) In vitro neurogenesis by progenitor cells isolated from the adult human hippocampus. Nat Med 6:271–277

27. Brustle O, Choudhary K, Karram K, Huttner A, Murray K, Dubois-Dalcq M, McKay RD (1998) Chimeric brains generated by intraventricular transplantation of fetal human brain cells into embryonic rats. Nat Biotechnol 16:1040–1044

28. Yu D, Neeley WL, Pritchard CD, Slotkin JR, Woodard EJ, Langer R, Teng YD (2009) Blockade of peroxynitrite-induced neural stem cell death in the acutely injured spinal cord by drug-releasing polymer. Stem Cells 27:1212–1222

29. Haragopal H, Yu D, Zeng X, Kim SW, Han IB, Ropper AE, Anderson JE, Teng YD (2015) Stemness enhancement of human neural stem cells following bone marrow MSC coculture. Cell Transplant 24:645–659

30. Wakeman DR, Redmond DE Jr, Dodiya HB, Sladek JR Jr, Leranth C, Teng YD, Samulski RJ, Snyder EY (2014) Human neural stem cells survive long term in the midbrain of dopamine-depleted monkeys after GDNF overexpression and project neurites toward an appropriate target. Stem Cells Transl Med 3:692–701

31. Snyder EY, Taylor RM, Wolfe JH (1995) Neural progenitor cell engraftment corrects lysosomal storage throughout the MPS VII mouse brain. Nature 374:367–370

32. Weiss S, Reynolds BA, Vescovi AL, Morshead C, Craig C, van der Kooy D (1996) Is there a neural stem cell in the mammalian forebrain. Trends Neurosci 19:387–393

33. Kim KS, Lee HJ, Jeong HS, Li J, Teng YD, Sidman RL, Snyder EY, Kim SU (2011) Self-renewal induced efficiently, safely, and effective therapeutically with one regulatable gene in a human somatic progenitor cell. Proc Natl Acad Sci U S A 108:4876–4881

34. Teng YD, Park KI, Larvik EB, Langer R, Snyder EY (2002) Stem cell culture: neural stem cells. In: Atala A, Lanzer RP (eds) Methods of tissue engineering. Academic Press, San Diego, CA, pp 421–437

35. Ryder EF, Snyder EY, Cepko CL (1990) Establishment and characterization of multipotent neural cell lines using retrovirus vector mediated oncogene transfer. J Neurobiol 21:356–375

36. Jia C, Yu D, Lamarre M, Leopold PL, Teng YD, Wang H (2014) Patterned electrospun nanofiber matrices via localized dissolution: potential for guided tissue formation. Adv Mater 26:8192–8197

37. Markowitz D, Goff S, Bank A (1988) Construction and use of a safe and efficient amphotropic packaging cell line. Virology 167:400–406

38. Teng YD, Kabatas S, Wakeman DR, Li J, Snyder EY, Sidman RL (2009) Chapter 16: Functional multipotency of neural stem cells and its therapeutic implications. In: Ulrich H (ed) Perspectives of stem cells: from tools for

studying mechanisms of neuronal differentiation towards therapy. Springer-Verlag, Inc., San Diego, CA, pp 255–270

39. Ropper AE, Zeng X, Haragopal H, Anderson JE, Aljuboori Z, Han Abd-El-Barr M, Lee HJ, Sidman RL, Snyder EY, Viapiano MS, Kim SU, Chi JH, Teng YD (2015) Targeted treatment of experimental spinal cord glioma with dual gene-engineered human neural stem cells. Neurosurgery 79(3):481–491

Chapter 12

Real-Time Dual MRI for Predicting and Subsequent Validation of Intra-Arterial Stem Cell Delivery to the Central Nervous System

Piotr Walczak and Miroslaw Janowski

Abstract

Stem cell therapy for neurological disorders reached a pivotal point when the efficacy of several cell types was demonstrated in small-animal models. Translation of stem cell therapy is contingent upon overcoming the challenge of effective cell delivery to the human brain, which has a volume of ~1000 times larger than that of the mouse. Intra-arterial (IA) injection can achieve a broad, global, but if needed also spatially targeted biodistribution; however, its utility has been limited by unpredictable cell destination and homing as dictated by the vascular territory, as well as by safety concerns.

We show here that high-speed MRI can be used to visualize the intravascular distribution of a super-paramagnetic iron oxide contrast agent and can thus be used to accurately predict the distribution of IA administered stem cells. Moreover, high-speed MRI enables the real-time visualization of cell homing, providing the opportunity for immediate intervention in the case of undesired biodistribution.

Key words Intra-arterial, Iron oxide, MRI, Real-time, Stroke

1 Introduction

There is growing interest in stem cell-based regenerative medicine, additionally fueled by the excellent results in small-animal models. Neurological disorders are of particular interest because of the paucity of radical solutions such as the organ transplantation. While the advances in stem cell biology are rapid, the appropriate delivery methods and studies on stem cell integration with a host brain are still in infancy. The major challenge is to deliver stem cells to the right, usually vast area in a minimally invasive fashion. The intraparenchymal route of cell delivery to the brain was dominating for years as it allows for unequivocal cell delivery to the brain. However, this method is relatively invasive as it requires brain puncture, and the volume of transplanted cell suspension may lead to the tear of the host tissue, which in turn may produce negative neurological consequences as well as trigger the attack by immune system [1]. The number of

Amit K. Srivastava et al. (eds.), *Stem Cell Technologies in Neuroscience*, Neuromethods, vol. 126,
DOI 10.1007/978-1-4939-7024-7_12, © Springer Science+Business Media LLC 2017

studies, which are using intravenous route, is skyrocketing, but there is agreement in scientific community that the cerebral engraftment is rather very low, thus the therapeutic effect is rather mediated through periphery [2]. Therefore, the finding of efficient route for cell delivery to the central nervous system (CNS) is very compelling. The body fluids are a very interesting route to the deliver cells to the brain in minimally invasive fashion. However, the fluid compartments are always associated with uncertainty about the final cell destinations; thus the effective uses of these routes require the development of techniques for precise and predictable delivery of stem cells.

The intra-arterial infusion of stem cells through endovascular catheter is on the raise. The growing interest is also related to the high efficacy of intra-arterial thrombectomy in stroke patients [3], opening up opportunity for adding stem cell infusion to the stroke treatment since the catheter is already in place. However, it is required to precisely navigate stem cell delivery within the cerebral vasculature, and this challenge might be addressed by the real-time imaging of stem cells.

The tracking of stem cells is already quite well established, and superparamagnetic iron oxide nanoparticles (SPION) are most frequently used for cell labeling [4–7]. This method has been also pursued in clinical studies [8]. However, until now, the imaging has been performed after the transplantation of stem cells [9–11], which is too late to actively navigate the process of cell infusion, which was a major motivating factor to develop a method to see cells flowing inside cerebral vasculature in real time. Moreover, that method can offer more utility for intra-arterial infusion as it allows even to predict the destination of intra-arterially derived stem cells [12].

2   Equipment, Materials, and Experimental Setup

2.1   MR Scanners, Coils, and Pulse Sequences

1. Clinical 3T scanner (Siemens Trio) was employed for large animal studies. A quadrature head coil has been used for brain imaging, and coil built into the bed of the scanner was used for imaging of the spine. The dynamic GE-EPI (TE = 36 ms, TR = 3000 ms, FOV = 1080, matrix = 128, acquisition time = 3 s, and 50–100 repetitions) was used to acquire real-time images. This pulse sequence is displayed at the screen in real time which allows for immediate feedback. The standard pre- and post-transplantation T2 and T2* images were acquired in addition to GE-EPI to provide high-resolution images and confirmation of cell destination.

2. A horizontal bore 7T or 11.7T scanner (Bruker) with a 15 mm planar surface coil was used for small animal imaging. Following acquisition of pre-transplantation baseline T2 and T2* images, real-time dual MRI of ferumoxytol perfusion and SPIO-labeled cell injections was performed using a standard echo

planar imaging (EPI) sequence with TE = 17 ms, TR = 2000 ms, FOV = 26 × 26 mm, matrix = 96 × 96, and acquisition time = 2 s with 200 repetitions. Standard T2-w and gradient-echo images were also acquired after labeled cell injection.

2.2 Stem Cells

We have used several types of stem cells at early passages (2–5) including:

1. Human mesenchymal stem cells (MSCs) (PT-2501; Lonza).
2. Human glial restricted precursors (hGRPs) (Q cells®; Q Therapeutics).
3. Mouse MSCs (mMSCs) that were derived from femurs through ultracentrifugation followed by selection based on adherence to plastic.
4. Porcine MSCs (pMSCs) that were isolated from iliac crest bone marrow following Ficoll density gradient centrifugation and selection using plastic adherence.
5. Mouse GRPs (mGRPs) were isolated according to a previously established protocol [13].

2.3 MRI Contrast Agents

The two types of SPION have been used.

1. The Molday ION-Rhodamine B (BioPAL) has been used for cell labeling. It is characterized by an excellent uptake by virtually any type of stem cells and high labeling efficiency.
2. Ferumoxytol (Feraheme®) is a SPION-based FDA-approved intravenous drug for the treatment of anemia, which has been also used frequently off-label as an intravenous contrast agent. Here, we perform intra-arterial infusion of ferumoxytol, and since it is merely local delivery, the systemic dose is negligible. We have also observed a fast clearance of ferumoxytol from cerebral arteries which indicates no risk for neurotoxicity. Thus, the ferumoxytol is an excellent candidate for the application in clinical studies.

2.4 Catheters

1. Large animals: 5-French angled glide catheter, 0.035-in. guide wire, 1.7-French microcatheter, 0.014-in. microwire
2. Small animals: carotid artery VAH-PU-C20 catheter (Instech Solomon Inc.) connected to #30 PTFE tubing

3 Methods

3.1 Cell Labeling

Labeling of cells with Molday ION-Rhodamine B (BioPAL, Inc.) was performed by overnight incubation with 20-µg Fe/mL of the contrast agent formulation. We have previously reported that this protocol does not negatively affect MSC (including viability) [14].

We have also successfully used this protocol in our previous studies with neural stem cells [15] or GRPs [1]. Prior to injection, cells were trypsinized, centrifuged at $786 \times g$ for 4 min, resuspended in 10-mM PBS, pH 7.4, filtered through a 70-µm nylon cell strainer (BD Falcon™), counted, checked for viability (viability at least 90% qualified cells for transplantation), and adjusted to $0.2–1 \times 10^6$ cells/mL.

3.2 Cannulation of Arteries Under Fluoroscopic Guidance in Large-Animal Models

Swine (male, weight = 30 kg, $n = 6$) were anesthetized with propofol (3 mg/kg/h), and a 5-French femoral arterial sheath was surgically introduced. A 5-French angled glide catheter was advanced over a 0.035-in. glide wire, and the right CCA was selectively catheterized using real-time fluoroscopy. Since pigs do not have an extracranial portion of the internal carotid artery (ICA), the catheter was advanced distally into the ascending pharyngeal artery proximal to the rete mirabile (route of major blood supply to the brain) using roadmap guidance.

For cerebral IA injection in dogs (male greyhound dogs, weight = 20–25 kg, $n = 2$), a 5-French femoral arterial sheath was surgically introduced, and a 5-French angled glide catheter was advanced over a 0.035-in. glide wire into the ICA under fluoroscopic guidance. A 1.7-French microcatheter was then advanced over a 0.014-in. microwire, and the origin of the MCA was selectively catheterized under roadmap guidance.

For spinal IA injection in dogs (male greyhound dogs, weight = 20–25 kg, $n = 2$), a 5-French femoral arterial sheath was surgically introduced. A 5-French reverse curve catheter was advanced over a 0.035-in. guide wire, and a limited spinal digital subtraction angiogram was performed on a single plane angiography unit (Artis, Siemens). Intersegmental arteries were selectively catheterized and injected with 2 mL of iodinated contrast (Omnipaque 300, GE Healthcare) to identify the origin of the Adamkiewicz artery (great radicular artery). Using the 5-French catheter as a guide catheter, a 1.7-French microcatheter was co-axially advanced over a 0.014-in. microwire to selectively catheterize the Adamkiewicz artery.

3.3 Stem Cell Transplantation in a Large-Animal Model

After fixation of catheter in desired position, the pigs were transferred from X-ray table to a clinical 3T MRI (TRIO, Siemens) scanner, and pre-transplantation T2 and SWI scans were acquired. Next, MSCs were injected IA at 5 mL/min while acquiring dynamic GE-EPI scans for real-time monitoring of cell inflow into the brain. The unique vascular anatomy of swine, with the network of arterioles on the skull base (rete mirabile), prohibits advancing an intravascular catheter into selective brain arteries. The nonselective catheterization results in transcatheter cell distribution to the entire hemisphere, which allows for determination of feasibility of real-time cell visualization of cell flow, but is insufficient for more precise cell targeting or to assess the predictability of cell distribution.

To address this issue, selective catheterization of cerebral arteries was performed in dogs, which are characterized by an intracranial vasculature that is more similar to that of humans. With the microcatheter at the origin of the MCA, we initially injected three separate short boluses of iron oxide-based MRI contrast agent: ferumoxytol (0.03 mg Fe/mL) [16], while acquiring dynamic GE-EPI (parameters are in a section below) to predict the area of cell distribution. Then, the MSCs were intra-arterially infused at the same speed. To confirm that the cerebral blood flow (CBF) was not altered by transplanted cells, three boluses of ferumoxytol followed cell delivery.

3.4 Middle Cerebral Artery Occlusion (MCAO) Stroke Model

Sprague-Dawley rats (male, 250 g, Harlan, $n = 4$) were anesthetized with 2% isoflurane, and an optic fiber was attached to the temporal bone after surgical cutdown to measure blood flow with laser Doppler (moorVMS-LDF, Moore Instruments). The internal carotid artery (ICA) was exposed with ligation of the extracranial branches [10]. After temporary closure of the ICA and the CCA by surgical clips, the arterial wall of the ECA was microdissected, and an Ethilon 4.0 suture with a silicon tip was manually advanced into the ICA. After removal of the clip, the suture was further advanced into the cerebral arteries until a decrease in cerebral blood flow of at least 70% (as measured by LDF) was reached. The suture was held in place for 45 min and then removed. The stump of the ECA was ligated, and the clip was removed from the CCA, restoring blood flow in the CCA and the ICA wounds, which were sutured and closed.

3.5 Lacunar Stroke Model

Wistar rats (male, 250 g, $n = 16$) were anesthetized with ketamine i.p. (90 mg/kg) and xylazine (10 mg/kg), immobilized in a stereotactic apparatus (Stoelting Inc.). A small burr hole was drilled in the skull over the right hemisphere. The needle, connected to a 10-μL syringe (Hamilton), was inserted into the right striatum (A = 0.0, L = 3.0, D = 3.5 mm). An injection of 3 μL of 5-mM ouabain (Sigma) in saline was given at a rate of 1 μL/min via a microinfusion pump (Stoelting Inc.) mounted on the stereotactic apparatus. The needle was then withdrawn, and the skin was closed with a nylon suture.

3.6 Carotid Artery Cannulation in Rats

For the lacunar stroke model, under general isoflurane anesthesia, the animals were positioned supine, and the CCA and ICA were dissected with the extracranial branches permanently ligated. For the MCAO stroke model, the surgical wound was reopened, and the carotid arteries were dissected within the postoperative scar, without approaching the PA or the stump of the ECA. The CCA was permanently ligated at the proximal site. For both stroke models, a catheter (VAH-PU-C20, Instech Solomon Inc.) connected to #30 PTFE tubing was introduced into the ICA and secured in place to the ICA and the CCA. Temporary sutures were applied to the surgical wound, and the animal was then transferred to the MRI scanner. After transplantation and imaging, the catheter was removed, and the surgical wound was permanently closed with sutures.

3.7 IA Stem Cell Delivery in the MCAO and Lacunar Stroke Model

SPIO-labeled hMSCs were injected IA in Sprague-Dawley rats at speeds 0.2 and 0.4 mL/min a day after induction of MCAO stroke. The MRI and bioluminescent imaging was performed during and after transplantation ($n = 2$).

SPIO-labeled MSCs were injected IA in Wistar rats ($n = 16$) at a speed 0.2 mL/min at various time points (1, 2, 3, and 7 days) after the induction of lacunar stroke [17]. Following the acquisition of baseline T2 and T2* sequences, SPIO-labeled hMSCs were loaded into the syringe attached to the catheter and infused. Dynamic acquisition of the MR images using GE-EPI was started 10 s before initiating the infusion of SPIO-labeled cells. We then evaluated IA delivery of SPIO-labeled small-size hGRPs (diameter ca. 13 µm) compared to that of larger-sized MSCs (diameter ca. 25 µm).

3.8 Bioluminescence Imaging (BLI)

Whole-body BLI was performed in rats to determine the biodistribution of transplanted Luc+ mGRPs. Anesthesia was induced with 5% isoflurane and maintained with 2% isoflurane/98% oxygen. Luciferin was administered i.p. at 150 mg/kg. Male rats (weighing ca. 250 g, $n = 2$) were placed inside a Spectrum/CT optical imager (PerkinElmer) and imaged every 5 min for a period of 30 min, with the acquisition time set to 1 min. The acquired images were processed using LivingImage® software (PerkinElmer).

3.9 Image Processing

OsiriX and Amira software were used for image visualization. Quantitative image analysis was achieved using MATLAB. A square-type waveform was assumed, representing the time course of injection. Reference waveforms were correlated with the pixel time course to form an array of p-values. p-Maps were calculated for each continuous dynamic acquisition. The p-values were normalized to the contralateral hemisphere (which did not receive injection) to mitigate the internal noise of each dataset.

Statistical analysis: Regression analysis is reported as Type III tests. The least square mean (LSM) values were used to detect differences from baseline and for comparison between means (PROC MIXED, SAS 9.4).

4 Results

4.1 Visualization of Intra-Arterial Cell Infusion Procedure to the Brain Using Real-Time MRI in a Porcine Model

Successful cannulation of the ascending pharyngeal artery with visualization of the rete mirabile was confirmed by x-ray fluoroscopy (Fig. 1a) with reference to the anatomy of cerebral vessels (Fig. 1b). Serially acquired GE-EPI images during the inflow of SPIO-labeled cells into the brain exhibited a gradual, focal reduction of pixel intensities (PI) over the period of cell injection.

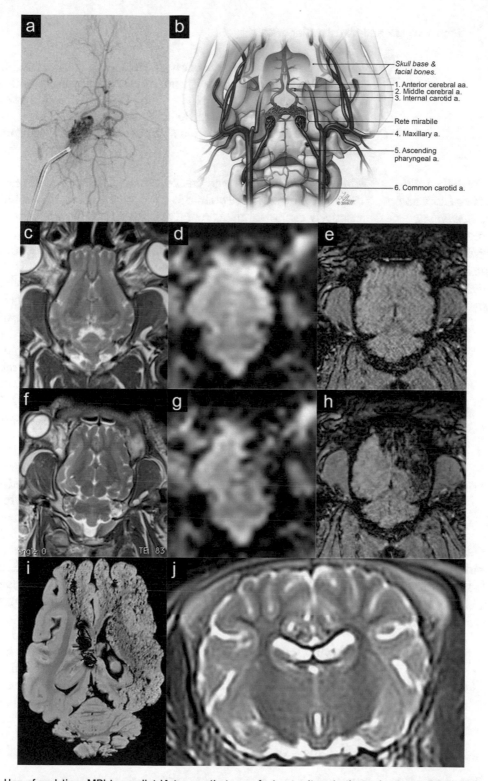

Fig. 1 Use of real-time MRI to predict IA transcatheter perfusion territory in the swine brain. (**a**) Placement of catheter in the ascending pharyngeal artery feeding the carotid rete of many intertwining arteries that supply cerebral blood flow. (**b**) Schematic of the cerebral vasculature in swine. T2-weighted MRI (**c, f**), real-time GE-EPI (**d, g**), and SWI MRI (**e, h**), pre- (**c–e**) and post-injection (**f–h**) of 5×10^6 MSCs at 1 mL/min. Cell engraftment occurred in nearly the entire hemisphere, with SWI and GE-EPI sequences being the most sensitive. (**i**) Ex vivo high-resolution MRI showing a punctate pattern of cell distribution within the brain. (**j**) T2-weighted image obtained after 1-week post-cell injection reveals normal brain anatomy with no signs of infarct/ischemia. Reproduced with permission from Ref. [12]

This corresponded to the process of MSC accumulation within the porcine brain (Fig. 1c–h), with the cell location subsequently confirmed by high resolution ex vivo MRI (Fig. 1i). Notably, follow-up T2 MRI scans performed the next day ($n = 3$) and after 1 week ($n = 1$) did not detect ischemia as a possible result of cell-induced microemboli (Fig. 1j).

4.2 Real-Time MRI Facilitates Prediction of Cell Destination with Cell Bio distribution Dependent on Catheter Tip Location in Canine Model

While pig experiments were instrumental in showing that real-time imaging of intra-arterial stem cell infusion is feasible, the inability to selectively cannulate the cerebral arteries due to the presence of the rete mirabile prevented investigation of spatial prediction of cell biodistribution. Thus, for this purpose, we used a canine model, as the dog cerebral vasculature is similar to that of the human.

A 0.3-mg/mL concentration of ferumoxytol was sufficient to detect a change in signal intensity on MRI. The observed perfusion area in the dog brain was found to correspond to the MCA territory (Fig. 2a, b). We then injected MSCs at the same speed and compared their spatial distribution to that of the ferumoxytol bolus and found an exact overlap. There was a moderate-to-strong correlation between the two p-maps derived from the ferumoxytol perfusion territory and cell engraftment territory ($r = 0.58$, $p < 0.05$). Since MSCs lodged in the brain capillaries, we confirmed that homing of cells to the brain does not compromise blood perfusion. Microscopic analysis of post-mortem brain tissue revealed SPIO-rhodamine-labeled red fluorescent cells in the targeted vascular territorial area, but not in the nontargeted areas (Fig. 2c). Quantitative analysis of the number of localized fluorescent cells revealed that the difference was statistically significant ($p < 0.05$).

4.3 Visualization of Intra-Arterial Delivery of MSCs to the Spinal Cord Using Real-Time MRI in a Canine Model

In contrast to the brain which due to magnetic homogeneity is relatively easy to image, spinal cord with close proximity of the bone, fluid, and muscle is a much more challenging organ for imaging. We wanted to test whether our real-time MRI of cell infusion is also feasible in thing challenging environment of spinal cord.

We demonstrated with MRI a broad distribution of infused cells within the lumbar and thoracic spinal cord. These images showed more cells engrafting to the regions of the proximal artery of Adamkiewicz (Fig. 3a, b; ROI 1 and 2) compared to the distal part (Fig. 3a, b; ROI 3 and 4). The dynamics of cell inflow are visualized in Fig. 3c. A rapid and extensive cell inflow to the segments proximal to the artery of Adamkiewicz was observed, with a slower and reduced cell inflow in distant regions of the spinal cord.

4.4 Near-Single-Cell Real-Time Visualization of Stem Cells in a Small Animal-Model of Stroke Under High-Field MRI

While the infusion of large MSCs results in spontaneous accumulation of MSCs in cerebral vessels, the small GRPs perfuse out of the cerebral vasculature, being stopped only when these cells are engineered for improved endothelial adhesion [10]. Here we tested with real-time MRI whether these distinct cell behaviors could be captured dynamically. As expected, large-size MSCs

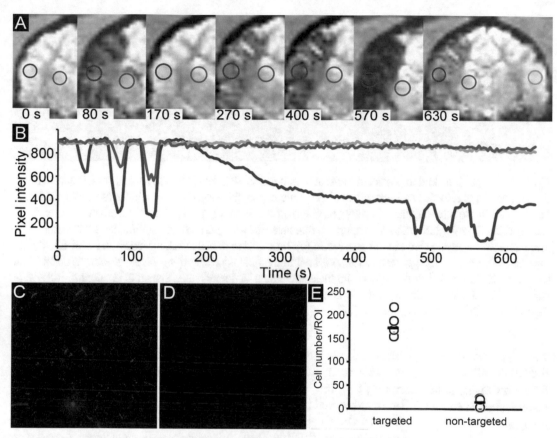

Fig. 2 Use of real-time dual MRI to predict IA transcatheter perfusion territory in dog brain. Three bolus injections of 300-μL Ferumoxytol® (0.3 mg Fe) were given 15 s apart at an injection speed of 1 mL/min within a 0–170 s interval, with subsequent clearance. Subsequent IA injection of SPIO-labeled cells infused at the interval between 170 and 400 s resulted in a gradual reduction of PI in the same region previously highlighted by ferumoxytol infusion. To confirm that the CBF was not compromised by the engrafted cells, ferumoxytol was injected a second time as three boluses after cell delivery, between the interval of 400–600 s, which resulted in clearance of ferumoxytol but not of transplanted cells (a). *Red* ROI represents the brain region of transplanted cell engraftment as predicted by ferumoxytol infusion. *Blue* ROI represents the region outside the targeted cell engraftment area, also as predicted by ferumoxytol infusion. *Green* ROI represents the contralateral hemisphere used as a control to validate the temporal stability of the image. Graph lines and ROIs are shown in corresponding colors (b). The darkening of MR image due to SPIO is illustrated at the graph as the drop of signal. The difference between the ipsilateral targeted hemisphere (c) and the contralateral nontargeted hemisphere (d) was also statistically significant on post-mortem evaluation (e, <0.0001). Localized cells are visible as red (rhodamine+). Reproduced with permission from Ref. [12]

gradually accumulated in the cerebral vasculature. Real-time MRI of cell infusion clearly demonstrated that small-size hGRPs were able to flow through the brain capillary bed, with only a few cell aggregates lodging within the vasculature. To avoid cell aggregation, the cells were infused immediately after preparation, and the real-time MRI was the only proof that cells flowed through the cerebral vessels.

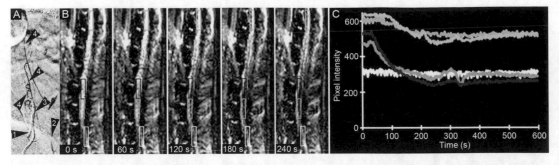

Fig. 3 Use of real-time MRI to predict IA transcatheter perfusion territory in the dog spinal cord. Digital subtraction angiogram of selective catheterization of the artery of Adamkiewicz. (1) 5-French reverse curve glide catheter located in the inter-segmental artery branching off the aorta, (2) 1.7-French microcatheter located in the Adamkiewicz artery (great radicular artery), (3) Adamkiewicz artery on the angiogram, and (4) anterior spinal artery directly supplying blood to the spinal cord (**a**). Time course of cell injection visualized by GE-EPI MRI (**b**). Several ROIs were drawn *(green* and *red,* proximal to the Adamkiewicz artery; *blue* and *orange,* distal to the artery, *white,* control below the artery) and their pixel intensity quantified as shown in (**c**). Control ROI *(white)* validates the stability of the GE-EPI MR image. *Graph lines* and ROIs are shown in corresponding colors. Reproduced with permission from Ref. [12]

4.5 Application of Real-Time MRI to Adjust the Speed of IA Cell Injection in Small Animals

Although we have previously shown that the speed of 0.2 mL/min is effective in intra-arterial delivery of stem cells to the intact rat brain [18], real-time MRI failed to show any inflow of cells to the brain subjected to MCAO when infused at the same speed (Fig. 4). Due to the small size of the surface coil, imaging of extracerebral tissues was unfeasible, but as transplanted cells were expressing luciferase [1, 19], we used whole-body bioluminescence imaging (BLI), to identify homing areas presumably outside the brain. BLI showed that cells homed to the ocular region, indicating that the injected cells were exclusively routed to the ophthalmic artery, an intracranial branch of the ICA, and failed to reach the cerebral arteries (Fig. 4f). We hypothesized that this might be due to a blood pressure imbalance as the carotid catheter is fully occlusive; there is no blood flow in the cannulated artery, and slow trans-transcatheter infusion is directed to the ophthalmic artery as a result of pressure resistance originating from the circle of Willis. Our results indicated that the blood pressure in the cerebral arteries might be higher in a stroke model, preventing the cells from being infused into the brain at low infusion speeds (0.2 mL/min). To address this question, we doubled the injection speed to 0.4 mL/min, and this yielded efficient targeting of the cells to the brain, as confirmed by MRI (Fig. 4). p-value maps revealed that in case of slow infusion rate, only 40 pixels changed to hypointense, and that was not significantly different from the baseline ($p = 0.35$). Fast cell infusion resulted in a change of 715 pixels, which was statistically significant compared to baseline ($p < 0.05$). There was also a difference between slow cell infusion and fast cell infusion images ($p < 0.05$). Consistently, the BLI signal that represented engrafted cells was detected from the brain region (Fig. 4k).

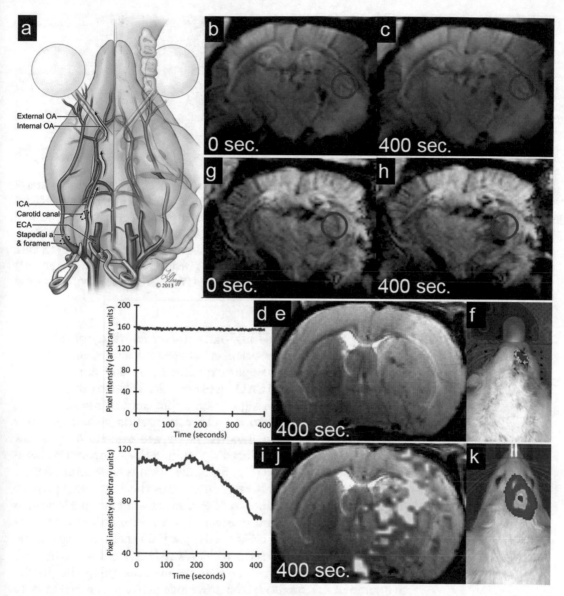

Fig. 4 Use of real-time MRI to adjust the cell injection speed during IA injection to ensure efficient stem cell delivery to the brain in rats. (**a**) Schematic of the cerebral circulation in the rat. (**b–e**) Injecting cells at 0.2 mL/ min does not result in cerebral engraftment, as no difference in pixel intensity (*red circle* = ROI, quantified in **d**) between the pre- (0 s) and post-injection (400 s) images was observed. (**f**) BLI of the same animal shows signal only above the ocular area. (**g–j**) In contrast, injecting cells at 0.4 mL/min results in effective cerebral engraftment, with a marked decrease in pixel intensity (**i**). (**k**) BLI confirms signal in the cerebral area without ocular localization. Color-coded subtraction images (**e, j**) highlight the difference in cell engraftment. Reproduced with permission from Ref. [12]

4.6 Real-Time MRI Reveals That Biodistribution of IA Injected Cells Depends on the Infusion Speed

We have above shown, in a large-animal model, that cell distribution territory depends on the catheter tip position and that this distribution can be predicted based on real-time MRI. Here, we investigated, in a small-animal model, whether the dependence of cell distribution on the infusion speed could also be monitored and

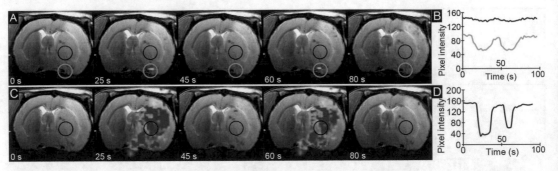

Fig. 5 Use of real-time MRI to predict IA transcatheter perfusion territory in the rat brain. Two bolus injections of 0.03-mg Fe were given 10 s apart at an injection speed of 0.2 mL/min (**a, b**) or 0.4 mL/min (**c, d**). Color-coded scale shows changes in signal intensity resulting from the transient ferumoxytol inflow. An inefficient cerebral perfusion is achieved at 0.2 mL/min speed (*black circle* = ROI, quantified in **b**), with a flow of contrast agent visible solely at the skull base near the ICA prior to the circle of Willis (*green circle* = ROI, quantified in **b**). In contrast, injecting ferumoxytol at 0.4 mL/min resulted in effective cerebral perfusion, with a 75% decrease in pixel intensity (**d**). Note that the ferumoxytol perfusion is transient and completely clears at 20 s after each injection. Reproduced with permission from Ref. [12]

predicted by imaging. Using transcatheter ferumoxytol injection, we observed a strong association between infusion speed and perfusion territory. In a single experiment, an infusion speed of 0.2 mL/min in the MCAO stroke model failed to route blood flow to the brain, with signal detectable only at the skull base, which can be assigned to the OA contribution area. Ineffective trans-catheter perfusion into the brain was evidenced by a low number of pixels with reduced signal intensity compared to baseline (283 pixels, $p = 0.4$) (Fig. 5). Similar results were found for the IA cell injections at this rate. Increasing the injection speed to 0.4 mL/min was sufficient to obtain successful brain perfusion, as evidenced by a significant number of pixels with reduced intensity compared to baseline (1720 pixels, $p < 0.05$) (Fig. 5). Again, the pixel mapping of perfusion results was consistent with the cell injection experiment described above, demonstrating the predictive value of this method. The difference between the two rates of infusion was statistically significant ($p < 0.05$). Importantly, IA-infused ferumoxytol completely clears from the brain vessels within seconds, enabling repetition of the procedure during the same imaging session until desired brain coverage is achieved.

4.7 Real-Time Imaging of Cell Inflow into the Lesion After IA Cell Delivery in a Lacunar Model of Stroke

The real-time MRI was capable of revealing the time course of cell distribution within the stroke lesion. PI graphs were generated to depict the time course for selected ROIs. No change of signal was observed in the contralateral hemisphere, demonstrating that the cells did not reenter the circulation with secondary lodging in the contralateral brain. Real-time MRI demonstrated that cells initially are located within the stroke periphery, with a delayed inflow into the core of the infarct. However, at the end of the infusion, the cell distribution within the overall infarcted area was quite homogeneous.

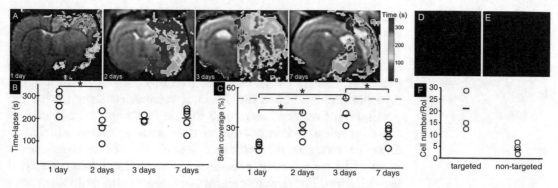

Fig. 6 Dependence of velocity and number of infused cells upon the interval between stroke induction and cell transplantation. Visualization of the velocity of cerebral cell inflow for different intervals between stroke induction and cell injection (*red*, high velocity; *blue*, low velocity) (**a**). Graph representing the dependence of cell inflow velocity on the time delay after stroke induction (**b**). Graph representing the percentage of entire brain cell engraftment as a function of the time delay after stroke induction (**c**). Note that the maximum brain engraftment that can be achieved was 50% (*dashed line*), as no contralateral flow was observed. Fluorescent images of tissue slices were used as a reference standard to validate targeted delivery. The difference between the ipsilateral targeted hemisphere (**d**), and the contralateral nontargeted hemisphere (**e**), was statistically significant on post-mortem images (**f**, $p < 0.001$). *Asterisks* represent statistical significance ($p < 0.05$). Reproduced with permission from Ref. [12]

By comparing different time intervals between stroke induction and cell injection, it became apparent that the temporal-spatial pattern of engraftment is interval dependent, with the slowest and lowest amount of cell homing at day 1 post-injection compared to later time points (Fig. 6). Histological analysis revealed that rhodamine-labeled cells indeed localized specifically in the regions previously shown as hypointense on MRI. Cell localization in regions that previously exhibited no MRI hypointensity was negligible (Fig. 6) ($p < 0.001$). Interestingly, real-time monitoring of cell infusion also indicated the existence of an extensive extracranial-intracranial anastomosis system in some animals, resulting in cells being carried from the brain to the muscles of the head.

5   Discussion

To the best of our knowledge, we are the first to show that high-speed MRI, based on a GE-EPI pulse sequence, enables monitoring of intra-arterial delivery of SPIO-labeled stem cells to the CNS in real time with sufficient sensitivity and high temporal resolution. The GE-EPI is frequently used for dynamic susceptibility contrast imaging in perfusion MRI with intravenous Magnevist® administration; thus most MR scanners are equipped with this pulse sequence [20]. The locoregional perfusion territory of individual arteries, although in general highly unpredictable, justifies the effort to predict cell biodistribution during catheter-based intra-arterial delivery. Real-time feedback about the temporal and spatial distribution of IA-injected ferumoxytol enables optimization of the position of the catheter and the infusion speed to ensure that cells reach the desired CNS regions with the highest possible precision. The inflow of cells and perfusion of the contrast agent into the brain depends on the local vascular resistance, with the possibility of reduced resistance in case of compromised cerebral autoregulation, as is encountered at some delay after cerebral ischemia. To overcome this resistance, it may be necessary to adjust the speed of IA cell injection according to the outcome of ferumoxytol pre-injection. The dynamic feature of GE-EPI also facilitates image processing and subtraction of background, enabling the creation of color-coded maps of cell biodistribution.

The real-time MRI was also capable of immediately detecting the potentially dangerous cell aggregates, which provides the opportunity for sudden cessation of cell infusion and the administration of a potentially de-aggregating therapy. The real-time monitoring of the procedure also answers questions about the lack of transplanted cells in the desired brain area, whether they were misrouted or did not adhere and stopped within the cerebral vessels.

As intra-arterial therapy of brain disorders has been found effective not only in animal models [21–23] but also in spontaneous animal disorders [24] as well as in a clinical trial [25], it is critical to deploy this therapy in a highly controlled manner, particularly as there is some variability in results, and not all authors have reported a prolonged positive effect [26]. It is expected that real-time monitoring of intra-arterial stem cell deliveries will further improve the therapeutic effect, as well as avoid complications in the clinical scenario. For example, it was shown that intra-arterial delivery of unmodified autologous bone marrow mononuclear cells is not superior to intravenous delivery [27], but, based on our results, the majority of small mononuclear cells probably went through the cerebral circulation unhindered, and more runs were necessary to accumulate in the brain parenchyma [28], which most probably was the reason for the lack of advantage of intra-arterial delivery.

Extensive evaluation of the safety of intra-arterial delivery was beyond the scope of this manuscript. However, we have previously performed detailed studies and established conditions for safe intracarotid infusion of stem cells, and we found that cell dose and the velocity of cell infusion are major determinants of safety [18, 29]. We postulate that monitoring of cell transplantation procedures in real-time under MRI will redefine the method by which safety is evaluated from group based to individual based. This personalized approach will most likely result in much improved safety of the procedure and increase the efficacy of transplanted cells.

Currently, placement of catheters in the cerebral arteries is routinely performed under the guidance of biplane fluoroscopy. For MR-guided transcatheter cell delivery, the subject must be transferred between the angiography suite and the MRI suite. However, there are rapid advances in the development of intravascular "MR microscopy," with MR coils built into the catheter [29]. With this approach, the entire procedure could possibly be performed in the future with real-time MRI as a "one-stop shop."

6 Conclusions

High-speed MRI based on a GE-EPI pulse sequence can be used for real-time monitoring of intra-arterial delivery of SPIO-labeled stem cells to the CNS. Furthermore, an SPIO-based contrast agent is capable of predicting stem cell destination, as well as verifying vessel patency after intra-arterial cell infusion. Thus, real-time MRI allows for highly precise intra-arterial infusion of stem cells to the CNS, which may facilitate the use of this minimally invasive route to deliver cells to the vast, but defined, brain regions, and is especially critical for large brains such as the human brain. Here, we have shown the feasibility of this technique in large animals using clinical MR scanners, making this approach fully clinically translatable.

Acknowledgments

We thank Mary McAllister for editorial assistance, and Lydia Gregg for preparing Figs. 1 and 4b.

References

1. Janowski M, Engels C, Gorelik M, Lyczek A, Bernard S, Bulte JW, Walczak P (2014) Survival of neural progenitors allografted into the CNS of immunocompetent recipients is highly dependent on transplantation site. Cell Transplant 23(2):253–262. doi:10.3727/096368912X661328

2. Janowski M, Wagner DC, Boltze J (2015) Stem cell-based tissue replacement after stroke: factual necessity or notorious fiction? Stroke 46(8):2354–2363. doi:10.1161/STROKEAHA.114.007803

3. Cougo-Pinto PT, Chandra RV, Simonsen CZ, Hirsch JA, Leslie-Mazwi T (2015) Intra-arterial therapy for acute ischemic stroke: a golden age. Curr Treat Options Neurol 17(7):360. doi:10.1007/s11940-015-0360-7

4. Walczak P, Kedziorek DA, Gilad AA, Barnett BP, Bulte JW (2007) Applicability and limitations of MR tracking of neural stem cells with asymmetric cell division and rapid turnover: the case of the shiverer dysmyelinated mouse

brain. Magn Reson Med 58(2):261–269. doi:10.1002/mrm.21280

5. Puppi J, Mitry RR, Modo M, Dhawan A, Raja K, Hughes RD (2011) Use of a clinically approved iron oxide MRI contrast agent to label human hepatocytes. Cell Transplant 20(6):963–975. doi:10.3727/096368910X543367

6. Puppi J, Modo M, Dhawan A, Lehec SC, Mitry RR, Hughes RD (2014) Ex vivo magnetic resonance imaging of transplanted hepatocytes in a rat model of acute liver failure. Cell Transplant 23(3):329–343. doi:10.3727/096368913X663596

7. Song M, Kim Y, Kim Y, Ryu S, Song I, Kim SU, Yoon BW (2009) MRI tracking of intravenously transplanted human neural stem cells in rat focal ischemia model. Neurosci Res 64(2):235–239. doi:10.1016/j.neures.2009.03.006

8. Bulte JW (2009) In vivo MRI cell tracking: clinical studies. AJR Am J Roentgenol 193(2):314–325. doi:10.2214/AJR.09.3107

9. Sheu AY, Zhang Z, Omary RA, Larson AC (2013) MRI-monitored transcatheter intra-arterial delivery of SPIO-labeled natural killer cells to hepatocellular carcinoma: preclinical studies in a rodent model. Invest Radiol 48(6):492–499. doi:10.1097/RLI.0b013e31827994e5

10. Gorelik M, Orukari I, Wang J, Galpoththawela S, Kim H, Levy M, Gilad AA, Bar-Shir A, Kerr DA, Levchenko A, Bulte JW, Walczak P (2012) Use of MR cell tracking to evaluate targeting of glial precursor cells to inflammatory tissue by exploiting the very late antigen-4 docking receptor. Radiology 265(1):175–185. doi:10.1148/radiol.12112212

11. Walczak P, Zhang J, Gilad AA, Kedziorek DA, Ruiz-Cabello J, Young RG, Pittenger MF, van Zijl PC, Huang J, Bulte JW (2008) Dual-modality monitoring of targeted intraarterial delivery of mesenchymal stem cells after transient ischemia. Stroke 39(5):1569–1574. doi:10.1161/STROKEAHA.107.502047

12. Walczak P, Wojtkiewicz J, Nowakowski A, Habich A, Holak P, Xu J, Adamiak Z, Chehade M, Pearl MS, Gailloud P, Lukomska B, Maksymowicz W, Bulte JW, Janowski M (2016) Real-time MRI for precise and predictable intra-arterial stem cell delivery to the central nervous system. J Cereb Blood Flow Metab, IN PRESS. doi:10.1177/0271678X16665853

13. Phillips AW, Falahati S, DeSilva R, Shats I, Marx J, Arauz E, Kerr DA, Rothstein JD, Johnston MV, Fatemi A (2012) Derivation of glial restricted precursors from E13 mice. J Vis Exp 64:3462. doi:10.3791/3462

14. Muhammad G, Jablonska A, Rose L, Walczak P, Janowski M (2015) Effect of MRI tags: SPIO nanoparticles and 19F nanoemulsion on various populations of mouse mesenchymal stem cells. Acta Neurobiol Exp (Wars) 75(2):144–159

15. Kedziorek DA, Muja N, Walczak P, Ruiz-Cabello J, Gilad AA, Jie CC, Bulte JW (2010) Gene expression profiling reveals early cellular responses to intracellular magnetic labeling with superparamagnetic iron oxide nanoparticles. Magn Reson Med 63(4):1031–1043. doi:10.1002/mrm.22290

16. Janowski M, Walczak P, Pearl MS (2015) Predicting and optimizing the territory of blood-brain barrier opening by superselective intra-arterial cerebral infusion under dynamic susceptibility contrast MRI guidance. J Cereb Blood Flow Metab. doi:10.1177/0271678X15615875

17. Janowski M, Gornicka-Pawlak E, Kozlowska H, Domanska-Janik K, Gielecki J, Lukomska B (2008) Structural and functional characteristic of a model for deep-seated lacunar infarct in rats. J Neurol Sci 273(1–2):40–48. doi:10.1016/j.jns.2008.06.019

18. Janowski M, Lyczek A, Engels C, Xu J, Lukomska B, Bulte JW, Walczak P (2013) Cell size and velocity of injection are major determinants of the safety of intracarotid stem cell transplantation. J Cereb Blood Flow Metab 33(6):921–927. doi:10.1038/jcbfm.2013.32

19. Gorelik M, Janowski M, Galpoththawela C, Rifkin R, Levy M, Lukomska B, Kerr DA, Bulte JW, Walczak P (2012) Noninvasive monitoring of immunosuppressive drug efficacy to prevent rejection of intracerebral glial precursor allografts. Cell Transplant 21(10):2149–2157. doi:10.3727/096368912X636911

20. Thornhill RE, Chen S, Rammo W, Mikulis DJ, Kassner A (2010) Contrast-enhanced MR imaging in acute ischemic stroke: T2* measures of blood-brain barrier permeability and their relationship to T1 estimates and hemorrhagic transformation. AJNR Am J Neuroradiol 31(6):1015–1022. doi:10.3174/ajnr.A2003

21. Yavagal DR, Lin B, Raval AP, Garza PS, Dong C, Zhao W, Rangel EB, McNiece I, Rundek T, Sacco RL, Perez-Pinzon M, Hare JM (2014) Efficacy and dose-dependent safety of intra-arterial delivery of mesenchymal stem cells in a rodent stroke model. PLoS One 9(5):e93735. doi:10.1371/journal.pone.0093735

22. Toyoshima A, Yasuhara T, Kameda M, Morimoto J, Takeuchi H, Wang F, Sasaki T, Sasada S, Shinko A, Wakamori T, Okazaki M, Kondo A, Agari T, Borlongan CV, Date I (2015) Intra-arterial transplantation of allogeneic mesenchymal stem cells mounts neuroprotective effects in a transient ischemic stroke model in rats: analyses of therapeutic time window and its mechanisms. PLoS One 10(6):e0127302. doi:10.1371/journal.pone.0127302

23. Du S, Guan J, Mao G, Liu Y, Ma S, Bao X, Gao J, Feng M, Li G, Ma W, Yang Y, Zhao RC, Wang R (2014) Intra-arterial delivery of human bone marrow mesenchymal stem cells is a safe and effective way to treat cerebral ischemia in rats. Cell Transplant 23(Suppl 1):S73–S82. doi:10.3727/096368914X685023

24. Zeira O, Asiag N, Aralla M, Ghezzi E, Pettinari L, Martinelli L, Zahirpour D, Dumas MP, Lupi D, Scaccia S, Konar M, Cantile C (2015) Adult autologous mesenchymal stem cells for the treatment of suspected non-infectious inflammatory diseases of the canine central nervous system: safety, feasibility and preliminary clinical findings. J Neuroinflammation 12:181. doi:10.1186/s12974-015-0402-9

25. Silachev DN, Plotnikov EY, Babenko VA, Danilina TI, Zorov LD, Pevzner IB, Zorov DB, Sukhikh GT (2015) Intra-arterial administration of multipotent mesenchymal stromal cells promotes functional recovery of the brain after traumatic brain injury. Bull Exp Biol Med 159(4):528–533. doi:10.1007/s10517-015-3009-3

26. Oh SH, Choi C, Chang DJ, Shin DA, Lee N, Jeon I, Sung JH, Lee H, Hong KS, Ko JJ, Song J (2015) Early neuroprotective effect with lack of long-term cell replacement effect on experimental stroke after intra-arterial transplantation of adipose-derived mesenchymal stromal cells. Cytotherapy 17(8):1090–1103. doi:10.1016/j.jcyt.2015.04.007

27. Yang B, Migliati E, Parsha K, Schaar K, Xi X, Aronowski J, Savitz SI (2013) Intra-arterial delivery is not superior to intravenous delivery of autologous bone marrow mononuclear cells in acute ischemic stroke. Stroke 44(12):3463–3472. doi:10.1161/STROKEAHA.111.000821

28. Brenneman M, Sharma S, Harting M, Strong R, Cox CS Jr, Aronowski J, Grotta JC, Savitz SI (2010) Autologous bone marrow mononuclear cells enhance recovery after acute ischemic stroke in young and middle-aged rats. J Cereb Blood Flow Metab 30(1):140–149. doi:10.1038/jcbfm.2009.198

29. Cui LL, Kerkela E, Bakreen A, Nitzsche F, Andrzejewska A, Nowakowski A, Janowski M, Walczak P, Boltze J, Lukomska B, Jolkkonen J (2015) The cerebral embolism evoked by intra-arterial delivery of allogeneic bone marrow mesenchymal stem cells in rats is related to cell dose and infusion velocity. Stem Cell Res Ther 6:11. doi:10.1186/scrt544

30. Erturk MA, El-Sharkawy AM, Bottomley PA (2012) Interventional loopless antenna at 7 T. Magn Reson Med 68(3):980–988. doi:10.1002/mrm.23280

Chapter 13

Standardized Cryopreservation of Stem Cells

Maria L. Thompson, Eric J. Kunkel, and Rolf O. Ehrhardt

Abstract

Successful commercial and clinical application of stem cells requires robust and practical cryopreservation protocols. Stem cells, particularly human embryonic stem cells and induced pluripotent stem cells, are notoriously sensitive to cryopreservation, requiring specialized protocols to maintain optimal cell viability and recovery. This chapter reviews the current state of stem cell cryopreservation and provides a clinically relevant, optimized method for controlled rate freezing and thawing of human stem cells that is reliable, inexpensive, and user friendly. This method successfully prepares stem cells for long-term cryogenic storage while ensuring maximal post-thaw cell viability.

Key words Cryopreservation, Stem cells, Freezing, Thawing, Standardization, Stem cell culture, Stem cell biology

---

1 Introduction

The application of stem cells to treat human disease is a long-standing medical practice. The ability of these cells to proliferate and differentiate into many different cell types means they can potentially restore other cells in the body lost to damage or disease. While this unique capability makes them highly valued, stem cells are also highly vulnerable to changes in their environment and need to be treated with care and precision. Changes in temperature, in particular, can affect stem cell function downstream, and proper quality controls should therefore be in place to avoid compromising their therapeutic value.

Cryopreservation of cells extends their stability, ensures their availability when needed, and maintains a diversity of cells lines and types in a minimal space. The process of cryopreserving stem cells involves harvesting the cells, adding a cryoprotectant, freezing the cells, and eventually thawing them. Since stem cells differentiate readily, each step should be carried out in a manner that optimizes cell viability while preserving an undifferentiated state. Successful cryopreservation depends on optimal freezing and thawing techniques and optimal frozen storage conditions.

Amit K. Srivastava et al. (eds.), *Stem Cell Technologies in Neuroscience*, Neuromethods, vol. 126,
DOI 10.1007/978-1-4939-7024-7_13, © Springer Science+Business Media LLC 2017

Several aspects of stem cell cryopreservation have changed over the last few years, most notably in the way in which the cells are suspended in freezing medium. Whereas, originally, delicate cells such as human embryonic stem cells (hESCs) were frozen after potentially harmful enzymatic or mechanical dispersion, over time it was recognized that this rather crude method often resulted in low post-thaw viabilities or even selection for aberrant genetic variations favoring rapid growth and survival [1–4]. More recent protocols recommend instead that more gentle enzymatic and nonenzymatic methods be used to disperse cells into near single-cell suspensions before freezing, resulting in a more homogenous distribution of cells in the freezing medium.

Though the importance of the thawing process to stem cell survival is frequently overlooked, in fact the cell thawing rate and technique can have a dramatic effect on downstream viability and recovery. Thawing should be as rapid as possible [5, 6] to minimize the growth of ice crystals inside the cells, though at the same time one should be cautious not to damage the cells through accidental overheating.

This chapter describes in detail how stem cell lines can be optimally frozen for long-term storage and optimally thawed for growth in a cell culture vessel.

2   Stem Cell Freezing

To prepare stem cells for the freezing process, one should first ensure they are healthy, actively growing, and at a high density, preferably just sub-confluent. As soon as these conditions are met in culture, a frozen bank of stem cells should be established in order to retain their unique characteristics and ability to differentiate [7]. Ideally, the cells should be maintained in antibiotic-free media, as there is evidence that commonly used antibiotics such as penicillin, streptomycin, and gentamycin have an adverse effect on the growth rate of some types of stem cells [8]. We recognize that antibiotics are often used for the sake of convenience, however, to avoid the risk of bacterial contamination. In such cases, it is recommended that stem cells be grown in the absence of antibiotics for at least 1 week or one passage prior to freezing. Stem cells should also be assayed for the presence of mycoplasma before freezing, as infection with mycoplasma will preclude their usefulness for downstream applications.

Dispersing cells into near single-cell suspensions before freezing can improve stem cell viability and recovery. For detaching stem cells from their substrate, Accutase is often used [9], particularly for embryonic stem cells (ESCs) and induced pluripotent stem cells (iPSCs); however, nonenzymatic methods such as PBS-EDTA may also be used. Once stem cells are detached, they are either allowed to settle out by gravity or pelleted by low speed centrifugation. They are then resuspended at a high density (1–5×10^6/ml), either in 1× freezing medium or in a relevant stem cell medium combined 1:1 with 2× freezing medium such as mFreSR (Stem Cell Technologies).

The freezing procedure for stem cells generally involves slow cooling in the presence of a cryoprotectant to protect the cells from the damaging effects of dehydration and intracellular ice formation. Wide ranges of cryoprotectant solutions are commercially available, and most of these have an advantage over homemade solutions in that they are sterile and endotoxin tested. DMSO is one of the most commonly used cryoprotectants, due to its efficiency and versatility. Like most cryoprotectants, however, DMSO is toxic to cells in a time- and concentration-dependent manner [10]. For this reason, it is removed or diluted as soon as possible after cell thawing.

The use of a controlled rate freezing technique that cools cells to −80 °C at a rate of at 1–2 °C/min is considered standard [11, 12]. To achieve this freezing rate, we are using the CoolCell® isopropanol-free passive freezing container that has achieved superior viability and recovery results (Fig. 1) compared with other common passive freezing methods [13]. Once cells have reached −80 °C (about 4 h when using a CoolCell® freezing container), it is recommended that the vials be transferred to liquid nitrogen as quickly as possible, both to minimize the temperature fluctuations where there is frequent user access to the −80 °C freezer and to better protect cells from degradation during long-term storage. For the actual transfer from the −80 °C freezer to liquid nitrogen, stem cells should be transported on dry ice or in a dry ice-based transport system to avoid temperature elevation.

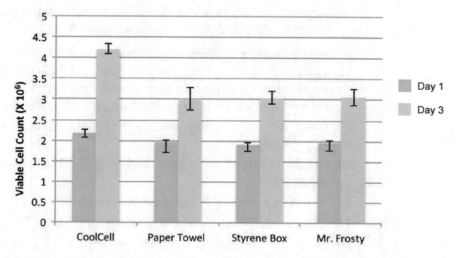

Fig. 1 Human embryonic stem cells, RC-10, were frozen using the technique indicated, thawed after 2 weeks in LN2, and counted immediately (*Day 1*) or after 3 days of growth (*Day 3*)

A logbook or bar code system should be used to record the names and locations of all stored cell lines.

2.1 Materials for Stem Cell Freezing

Note: all reagents should be cell culture grade unless otherwise noted.

1. 70% ethanol/water solution for disinfection. Water for solution should be sterile-filtered, suitable for cell culture (*see* **Note 1**).

2. Cell culture: actively growing pluripotent stem cells plated on 6-well dish (each well near confluency can accommodate up to five million cells per well).

3. 0.5 mM PBS-EDTA, pH 7.4.

4. Phosphate-buffered saline, pH 7.4, containing 0.5 mM EDTA (PBS-EDTA).

5. Standard stem cell medium: mTeSR1 (StemCell Technologies) is the most widely used medium; however, there are several well-functioning alternatives that are defined and/or xenobiotic-free (newer-generation low-protein media may not function as well in this protocol):

   (a) PeproGrow-hESC (BSA-containing defined medium; PeproTech).

   (b) StemPro® NSC SFM Complete Medium (serum free, ThermoFisher).

   (c) PsGro hESC/IPSC Growth Medium (human serum albumin, containing xenobiotic-free medium; System Bioscience, *http://www.systembio.com/*)

6. 1× or 2× freezing medium (*see* recipe, **Note 2**) in 15-ml tubes.

7. Liquid N_2.

8. Spray bottle for 70% ethanol.

9. Cryogenic vials, leak-free (liquid N_2 tested), sterile (TruCool brand recommended; *http://biocision.com/*).

10. Indelible marker, resistant to cold, water, and ethanol.

11. CoolBox™ 2XT workstation with CoolRack® CFT30 thermo-conductive cryogenic tube module and CoolRack® 15-ml conical tube module (Brooks Automation, Inc *http://brooks.com/*).

12. CoolCell® cell freezing container (Brooks Automation, Inc *http://brooks.com/*).

13. Serological pipet assortment, including 5-ml glass serological pipets.

14. Cell scraper (Sarstedt brand recommended).

15. 15-ml conical-bottom centrifuge tubes, sterile (e.g., BD Falcon).

16. Beckman refrigerated centrifuge with TH-4 rotor (or equivalent).

17. Liquid N_2 tank, preferably vapor tank for vapor phase storage.

2.2   Method for Stem Cell Freezing

Note: The use of sterile technique throughout these procedures is assumed.

Prepare equipment:

1. Label sterile cryogenic vials using an indelible marker with name of cell line, medium, and date. Alternatively, cryogenic labels can be made in the laboratory using ultra-low-temperature compatible labels with a thermal transfer printer [Zebra printer such as 3844Z (http://www.zebra.com/) with cryogenic labels (thermal-competent labels: 1-in. label plus 12-mm circle) from GA International Inc. (http://ga-international.com/)] and standard word processing software.

2. Wipe surfaces of all CoolRack® modules to be used in procedure with 70% ethanol solution.

3. Place the vials in CoolRack® CFT30 module (or other thermos-conductive rack) inside a CoolBox™ 2XT workstation that has XT Cooling Cores that have been prechilled. Harvest actively growing stem cell line using favored methodology. Here we recommend a gentle nonenzymatic method using PBS-EDTA. Please note that while Accutase will also produce similar small clumps or single cells, this enzymatic treatment or Dispase, unlike PBS/EDTA, does greatly benefit from the use of Y-27632 during the thaw phase (*see* **Note 3**).

4. Remove 6-well dish from humidified incubator and rinse each well of the 6-welldish to be frozen with 1–2 ml of PBS-EDTA, aspirate, and replace with 1 ml of PBS-EDTA.

5. Replace dish in incubator and wait 3–6 min.

6. Observe dish at 3-min time point, and, if the colonies appear to have a Swiss-cheese appearance, then they are ready to harvest.

7. Remove the PBS-EDTA, and, using a 5-ml pipet, gently squirt 3–5 ml of standard stem cell medium on each well while moving your pipet tip in a circular fashion.

   This should wash off all colonies in small clumps. DO NOT TRITURATE. You may also remove the PBS-EDTA, replace with medium (2–3 ml), and gently scrape off the cells with two to three swipes of a cell scraper. Proceed to step 8 . If the cells detach from the well during the incubation with PBS-EDTA, then dilute with three parts stem cell medium. Proceed to step 8 .

8. Using a 5-ml pipet, slowly transfer cell suspension (DO NOT TRITURATE) to a 15-ml conical tube, and place into a prechilled CoolRack® 15-ml module in the CoolBox™ 2XT workstation.

   You can repeat steps 1–6 for multiple dishes during step 9 below.

9. If you choose to use 2× freezing medium, proceed to **step 10**. Otherwise, centrifuge your sample at 120 to 150 × *g* (500–700 rpm in Beckman TH-4 rotor) at 4 °C with no brake.

Remove supernatant and replace with appropriate volume of prechilled 1× freezing medium. Place into CoolRack® 15-ml module and proceed to **step 10**.

10. After 5–10 min, gently remove tube from CoolRack® module and observe the settling of the small clumps into a loose cell pellet.

11. Remove the supernatant, reducing the sample volume to 0.5 ml per cryogenic vial.
From each well, you can freeze one to three vials. Each vial requires 0.5 ml of medium and 0.5 ml of 2× freezing medium. If you harvested six wells, then reduce volume to 3 ml of stem cell medium. The volume of the reduced sample should be checked using a glass serological pipet or individual tubes that have been calibrated for volume to ensure accurate addition of the 2× freezing medium.

12. Gradually add an equal volume of the prechilled 2× freezing medium to the cell pellet to achieve a final cell concentration of $1–5 \times 10^6$ cells/ml (*see* **Note 4**).

13. Close the tube and resuspend the cells by slow inversion.
When freezing small amounts of cells, do not invert, but gently tap the tube and use the tip of the 5-ml glass pipet to gentle mix the sample.

14. Transfer 1 ml of resuspended cells into each labeled cryogenic vials in the chilled
CoolRack® CFT30. When all the vials are filled, immediately place them in a CoolCell® cell freezing container and place in a −80 °C freezer for at least 4 h. All spaces in the CoolCell® cell freezing container must be filled with samples or "dummy" vials filled with the same amount of cryopreservative solution only to ensure proper temperature distribution during the freezing process.

15. For short-term storage (3 days or less), leave the cells at −80 °C (see Critical Parameters). For long-term storage, transfer the vials into a liquid nitrogen freezer after they have been in the −80 °C freezer for at least 4 h.

16. Record the appropriate information about the cells in the repository. Records should include all of the following information: culture identity, passage number or population doubling level, date frozen, freezing medium and method used, number of cells per vial, total number of vials initially frozen and number remaining, their locations, their expected viability, and results of all quality-control tests performed.

3 Thawing Stem Cells

As with freezing, the thawing step is an important part of the cryopreservation process and must be carried out properly to protect stem cell viability and function. Mechanical and chemical stresses that build up during the freezing and thawing process can trigger apoptotic pathways, resulting in post-thaw cell death. For this reason, apoptotic inhibitors, such as caspases and ROCK inhibitors, are sometimes added post-thaw to mitigate the effects of this stress, resulting in higher cell survival rates [14, 15].

Rapid thawing is recommended both to avoid damage from ice recrystallization and to transfer the resuscitated cells out of their freezing medium as soon as possible to avoid the toxic effects of cryoprotectants. For the same reasons, one should also avoid repeated freeze-thaw cycles, as exposing cryopreserved stem cells to elevated temperatures for even a short period of time can lower viability and recovery rates [15]. Rapid thawing of frozen cell samples has been shown to improve viability and recovery of many cell types, including hematopoietic stem cells [5] and mesenchymal stem cells [16].

Standardizing cryopreservation workflows reduces variability in research results and improves reproducibility, both of which are desirable goals for therapeutic cell types, especially in a GMP-regulated environment. Current accepted methodology for thawing cells is *not* standardized, but generally involves manually thawing cells in a 37 °C or warmer water bath, a process which relies on the judgment of the individual researcher, and is therefore subjective, and which can also introduce the risk of contamination from the surrounding water. Water baths can also be a problem in the clinical or preclinical facility, because FDA regulations require that they be constantly monitored for temperature fluctuations and sterility, a time-consuming process.

In this protocol, therefore, we will take advantage of the ThawSTAR™ automated cell thawing system, a dry heat system which addresses the emerging need to standardize cell thawing. The ThawSTAR™ system has a number of safety features incorporated into its design that enhance reproducibility (Fig. 2) while eliminating the potential for water-introduced contamination.

3.1 Materials for Stem Cell Thawing

1. ThawSTAR™ Automated Vial Thawing System (MedCision Inc http://medcision.com/).

2. ThawSTAR™ CFT Transporter (MedCision Inc).

3. Frozen cryogenic vial containing cells (see Freezing Stem Cells protocol).

4. 70% ethanol.

5. Thawing/culture medium (identical to that used for growth of cells).

6. Dry ice.

7. 15 ml tube with cap, sterile.

8. Eppendorf centrifuge 5810 (or equivalent).

9. Appropriate cell culture vessel.

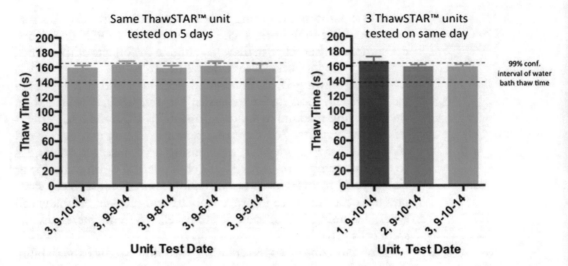

Fig. 2 ThawSTAR™ automated cell thawing system has a highly reproducible thawing profile. *Left panel*: the average time for the same ThawSTAR™ system to thaw >5 frozen vials each day for 5 days. Right panel: the average time for three different ThawSTAR™ systems was measured using >5 frozen vials per unit on 1 day. No significant differences were identified for either scenario at $p < 0.05$ (2-way ANOVA with post hoc Sidak test). For comparison, six vials were thawed in a 37 °C water bath. The average thaw time in the water bath was 151 s, with a 99% confidence interval of 139–164 s (range shown as *dotted lines*)

10. CoolRack® 15 ml thermo-conductive tube module (Brooks Automation, Inc).

Note: The use of sterile technique throughout these procedures is critical. Be sure to have full protective equipment while handling liquid nitrogen and dry ice.

3.2 Method for Stem Cell Thawing

1. Prepare equipment:
   Label new cell culture vessel with information about the cell type as needed. Power up the ThawSTAR™ system and allow 3–5 min for warm up.

2. Prepare a 15 ml conical tube with 9 ml of appropriate recovery medium and place in CoolRack® 15 ml module equilibrated at room temperature. This tube will be used to wash out the DMSO and other cryoprotective agents.
   Thaw cells:

3. Retrieve cryogenic vial containing frozen cells from storage area, and place immediately in a precooled ThawSTAR™ CFT Transporter. Allow at least 10 min for equilibration to <−70 °C before proceeding to the next step.

4. Insert the frozen cryogenic vial into the ThawSTAR™ cell thawing system. Push down on the vial to automatically initiate the thaw cycle. Retrieve thawed vial immediately upon completion as indicated by the audible and visual alert signal.

5. Wipe the outside of the vial with 70%ethanol to prevent contamination and place into a prechilled CoolRack CFT30, using an ice-free cooling base if needed to maintain <4 °C, inside a sterilized tissue culture hood.
   Some cells and freezing medium are temperature sensitive, so this step is critically cell type dependent.

6. Open vial and gently pipet out cells using a transfer pipette with a large orifice to avoid damaging cells. Transfer the cells to a sterile labeled 15 ml conical tube with 9 ml of room temperature thawing medium (original culture medium containing at least 10% fetal bovine serum or equivalent).

7. Cap the 15 ml tube, and centrifuge for 5 min at $300 \times g$.
   To avoid injury to the cells, the cell pellet should be loose and easy to resuspend with very little pipetting velocity.

8. Discard supernatant, resuspend cells in culture medium to desired concentration, and transfer to a tissue culture flask or other desired culture vessel.

9. A small sample may be taken from this cell suspension for immediate cell counting and viability analysis or phenotyping using, for example, a flow cytometer.
   Specific protocol optimization may include modification to substitute post-thaw centrifugation with simple post-thaw dilution.

4   Notes

1. 70% ethanol for cell culture: Ethanol for disinfection need not be high grade, and standard denatured ethanol is often used. Please remember, however, that ethanol/water mixtures do not have additive volumes; measure out 700 ml ethanol, and then bring to 1 L volume with water.
   Keep in mind when using diluted ethanol that you will have gradual evaporative loss, resulting in a gradual decrease in ethanol concentration. Some researchers use 95% ethanol solution instead, which is the highest concentration of ethanol you can get by distillation. This can be purchased directly from any reputable supplier.

2. Freezing medium: (1×) is commonly composed of a base medium (DMEM, or relevant stem cell media) containing a cryoprotectant such as dimethyl sulfoxide (10% v/v DMSO) along with a protein source such as fetal bovine serum (FBS; 10% to 20% v/v), knockout serum replacer (KoSR, 20% v/v final), bovine serum albumin (5% final), or human serum albumin (5% to 10% final).

3. Accutase (Life Technologies) may be used as an alternative to this nonenzymatic method; however, this will produce very small clumps or a near single-cell suspension, and settling out

of cells may not be a viable part of the protocol. Further, we recommend the addition of 2 μM Y-27632 (e.g., Sigma-Aldrich) upon passaging with enzyme, to improve cell viability. This compound can also be included in the freeze/recover medium for cells processed using Accutase.

4. Optimal concentration of frozen cells will vary by cell type and can be anywhere from 5 to 30×10^6 cells per ml. Please predetermine the concentration best suited to your cell type; this information is generally available from the supplier.

5   Conclusions

Cryopreservation is an important part of the process of producing stem cells and preserving their function. Given that the perceived value of stem cells is derived from their therapeutic potential, it is essential that frozen stocks are not only readily available, but that these cryopreserved stem cells are of the highest achievable integrity. The ongoing move toward standardized technologies and more reproducible results is a necessary one if stem cell-based therapies are to succeed.

Historically, the process of freezing viable stem cells was carried out by costly specialized equipment such as electronic programmable freezers. Now a more economical alternative is available in the CoolCell® container, an alcohol-free passive cell freezing container that freezes cells at a rate of 1 °C/min in a standardized and reproducible manner. The CoolCell® cell freezing container has already been adopted for work in a GMP facility, as well as for clinical trial use [17]. Correspondingly, the ThawSTAR™ automated cell thawing system brings standardization to cell thawing. This water-free system has a highly reproducible thermal profile and a rapid, controlled thawing rate designed to be nearly identical to that of a 37 °C water bath. It can be used within the sterile environment of a tissue culture hood and has been used to successfully cryopreserve hematopoietic stem cells, as well as other cell types [18].

In conclusion, this protocol demonstrates a practical and reliable method for optimal freezing and thawing of stem cells for long-term storage.

References

1. Peterson SE, Loring JF (2014) Genomic instability in pluripotent stem cells: implications for clinical applications. J Biol Chem 289(8):4578–4584

2. Martin-Ibanez R et al (2008) Novel cryopreservation method for dissociated human embryonic stem cells in the presence of a ROCK inhibitor. Hum Reprod 23:2744–2754

3. Li X et al (2008) The ROCK inhibitor Y-27632 enhances the survival rate of human embryonic stem cells following cryopreservation. Stem Cells Dev 17:1079–1085

4. Li X et al (2009) ROCK inhibitor improves survival of cryopreserved serum/feeder-free single human embryonic stem cells. Hum Reprod 24:580–589

5. Berz D, McCormack EM, Winer ES (2007) Cryopreservation of hematopoietic AU:5 stem cells. Am J Hematol 82:463–472
6. Chua KJ, Chou SK (2009) On the study of the freeze-thaw thermal process of a biological system. Appl Therm Eng 29:3696–3709
7. Lund RJ, Narva E, Lahesmaa R (2012) Genetic and epigenetic stability of human pluripotent stem cells. Nat Rev Genet 13:732–744
8. Cohen S et al (2006) Antibiotics reduce the growth rate and differentiation of embryonic stem cell cultures. Tissue Eng 12(7):2025–2030
9. Bajpai R et al (2008) Efficient propagation of single cells: accutase-dissociated human embryonic stem cells. Mol Reprod Dev 75(5):818–827
10. Morris C et al (2014) Should the standard dimethyl sulfoxide concentration be reduced? Results of a European group for blood and marrow Transplantation prospective non-interventional study on usage and side effects of dimethyl sulfoxide. Transfusion 54:2514–2522
11. Hunt CJ, Armitage SE, Pegg DE (2003) Cryopreservation of umbilical cord blood: tolerance of CD34+ cells to multimolar dimethyl sulphoxide and the effect of cooling rate on the recovery after freezing and thawing. Cryobiology 46:76–87
12. Naaldjik Y et al (2012) Effect of different freezing rates during cryopreservation of rat mesenchymal stem cells using combinations of hydroxyethyl starch and dimethylsulfoxide. BMC Biotechnol 12:49
13. Yokohama WM, Thompson ML, Ehrhardt RO (2012) Cryopreservation and thawing of cells. Current Protoc Immunol Appendix 3:3G. doi:10.1002/0471142735.ima03gs99
14. Watanabe K et al (2007) A ROCK inhibitor permits survival of dissociated human embryonic stem cells. Nat Biotechnol 25:681–686
15. Bissoyi A et al (2014) Targeting cryopreservation-induced cell death: a review. Biopreserv Biobank 12(1):23–34
16. Norkus M et al (2013) The effect of temperature elevation on cryopreserved mesenchymal stem cells. Cryo Letters 34(4):349–359
17. Foussat A et al (2014) Effective cryopreservation and recovery of human regulatory T-cells. BioProcess Int 12(S3):34–38
18. Stone M., et al. (2015) Maximizing PMBC recovery and viability: a method to optimize and streamline peripheral blood mononuclear cell isolation, Cryopreservation, and thawing. Bioprocess Int epub April 2015

Chapter 14

Isolation and Characterization of Extracellular Vesicles in Stem Cell-Related Studies

Zezhou Zhao, Dillon C. Muth, Vasiliki Mahairaki, Linzhao Cheng, and Kenneth W. Witwer

Abstract

Extracellular vesicles (EVs) are small derivatives of the cell that may mediate at least in part the effects of stem cells in vitro and in vivo. In light of strong interest in stem cell-based therapies, EVs are particularly attractive as therapeutic entities because they do not replicate and may have other safety advantages over whole stem cells. Here, we examine the current literature on stem cell-derived EVs, finding that differential centrifugation remains the most widely used central purification method for EVs. We then present a common variation of ultracentrifugation-based EV enrichment, followed by a protocol for determination of particle size and concentration by single particle tracking. Notes are provided to introduce technical variations and important considerations. Since rapid shifts in technique usage are expected to occur over the next several years, we briefly review new approaches to isolation and characterization.

Key words Extracellular vesicle, Exosome, Microvesicle, Ultracentrifugation, Nanoparticle tracking analysis, Stem cell, Mesenchymal stromal cell, Standardization

---

1 Introduction and Survey of Current Techniques

Extracellular vesicles (EVs) are bilayer membrane-enclosed particles released by most cell types [1, 2] and comprised of lipids, proteins, nucleic acids, and more. Ranging in size from tens to hundreds of nanometers in diameter—or even larger in the case of large oncosomes that are derived from some cancer cells [3]—EVs encompass a diversity of vesicle subtypes that are only beginning to be understood. Physiological roles for EVs in processes including coagulation and bone calcification have been suspected or known for decades [4–8], but widespread investigation of EVs has intensified only in recent years [9, 10]. In addition to a "garbage disposal" function, some EVs may participate actively in paracrine signaling, transmitting information to distant cells [11–16].

In stem cell research, it has been known for some time that stem cell conditioned medium (CM) could confer effects similar to

Amit K. Srivastava et al. (eds.), *Stem Cell Technologies in Neuroscience*, Neuromethods, vol. 126,
DOI 10.1007/978-1-4939-7024-7_14, © Springer Science+Business Media LLC 2017

those of the stem cells themselves [17]; it is now clear that EVs are the active component at least in some biological settings. The therapeutic and largely paracrine effects of, e.g., mesenchymal stromal cells (MSCs) are mediated in part by EVs [18, 19]. The discovery of these EV-mediated effects of stem cells has profound implications for the development of therapeutics (see [20] and references therein), not least because the use of EVs may sidestep some safety and regulatory issues. The EV does not have a nucleus and cannot proliferate. Also, when injected, smaller EVs may be less likely than cells to produce a pulmonary first pass effect [21].

How important are isolation and characterization techniques in stem cell-related EV research? The answer depends very much on the research question(s). For mechanism-focused basic research into the biological effects of naturally produced EVs, only relatively pure and well-characterized EV populations will suffice [10, 22]. Purity must be verifiable, and interesting features of the EVs must be manipulable. On the other hand, these considerations may take second place [20] if the subjects of research are EVs or EV-like bodies that are artificially produced, even extruded from cells. For example, EV "mimetics" produced from cells by microfluidic methods have been reported to deliver cargo as well as EVs produced biologically from the same cells at much lower concentration and greater cost [23]. If a relatively impure "releasate" or cellular extrusion mixture achieves a desired effect, a tedious and expensive purification procedure may not be essential at the beginning. However, translation requires viability from a regulatory perspective. Modern medicine based on good manufacturing practice (GMP) production requires a highly reproducible as well as safe procedure as well as knowledge of the active component. Therefore, methodologies of EV purifications and quantification are important for improvement of production and scale-up for future clinical applications. In general, purity still matters as we strive to understand mode of action and develop a reproducible and robust production procedure.

1.1 The State of the Field: Purification of Stem Cell-Derived EVs

In preparing this work, and in order to focus our attention on relevant issues, we sought to determine how EVs are currently studied in the stem cell field. We posed three main questions: What cellular sources are used? What vesicles are examined? What methods are followed to obtain EVs? We used PubMed and Boolean operators to search for articles on stem cells and extracellular vesicles, exosomes, or microvesicles that had been published through October 2015. Two hundred and sixty-eight articles were returned. Entries were reviewed to exclude the following: (1) studies that did not examine stem cell-derived EVs (such as articles on stem cells exposed to EVs from non-stem cell sources or investigations of the intracellular RNA processing complex known as the "exosome"); (2) articles that did not include primary data, such as review, hypothesis, and methods; (3) articles not written in the English language, since we did not have resources to engage expert translation; and (4) articles not available freely or through our institute's library subscriptions. We also attempted to exclude pieces published by recognized predatory publishers, since these publishers may have substandard review processes [24]. Seventy-eight articles on stem cell-derived EVs remained after application of these filters [18, 19, 25–99].

1.1.1   Cellular Source

Overwhelmingly, MSCs that can be easily established and expanded have been the most intensively studied source of EVs as reported in the existing literature. At least 51 of the 78 studies we reviewed derived EVs from MSCs. Only three studies involved iPSCs; in two of these, the iPSCs were used to derive MSC populations (studies published since the review period, e.g., [100], are not included here). Embryonic stem cells were the subject of six articles. In the remaining studies, the stem cell populations were referred to as progenitor cells of cardiac, endothelial, or neural origin; hematopoietic stem cells; and various cancer stem cells.

1.1.2   EV Subtypes and Nomenclature

Among the group of stem cell EV articles we considered, the majority, 46, referred to the studied population of EVs mainly as "exosomes," either alone (39 articles) or in combination with other terms (7 articles). Thirty used the term "microvesicle" (including the variants "micro-vesicle" and "microvessel"). Only four articles consistently used the term EV, although others indicated diversity with descriptions such as "extracellular membrane particles," "membrane vesicles," and "exosome-enriched extracellular particles" or even "microvesicles, termed exosomes." One article simply called the particles "vesicles." Just as in the wider EV literature, the same terms are often used to refer to different types of vesicles. Along with the International Society for Extracellular Vesicles, we recommend the use of the agnostic term "extracellular vesicle," EV, because vesicle subcategories have not yet been defined consistently by biogenesis, size, surface markers, or other characteristics [101]. If terms such as "exosome" or "microvesicle" are used instead, they must be defined at the outset, and rigorously collected evidence for these categorizations should be provided.

1.1.3   EV Isolation

The most commonly used technique of EV isolation—and truly the mainstay of modern EV research—remains stepped or differential centrifugation: EV-containing fluids are spun at successively higher speeds to enrich successively smaller particles. Perhaps the clearest exposition of this and related methods is a highly cited 2006 work from Clotilde Théry and colleagues [102]. This methods paper details and explains the process for several starting materials, including blood products and cell culture conditioned medium. For the most part, EV centrifugation protocols follow this procedure exactly or use slight modifications.

The protocol as described by Théry et al. [102] begins with a low-speed spin [e.g., 300 relative centrifugal force (rcf, or "× g") for 10 min]. The purpose of this initial step is to remove cells and large debris. The supernatant is then centrifuged at 2000 rcf for 10 min to deplete additional debris and large bodies. A 30-min, 10,000 rcf spin follows, depleting the fluid of debris and relatively large particles that are sometimes referred to as "microvesicles." Next, the supernatant is ultracentrifuged at 100,000 rcf or more to pellet small EVs, commonly termed "exosomes." This population is typically more homogeneous than the particles in the 10,000 × g pellet. One or more wash and re-pelleting steps may be done to deplete contaminants (such as protein aggregates) from the ultracentrifuge pellet. The resuspended pellet can then be used directly or subjected to further rounds of purification, for example, by velocity or flotation gradients [103–106].

1.1.4 What Protocols Are in Current Use in Stem Cell EV Research?

As in EV research overall, stepped centrifugation is the mainstay of current stem cell EV studies. At least 65 of the 78 articles we evaluated used stepped centrifugation in some capacity. For most of these, ultracentrifugation was the main isolation process. Twenty-six studies used variations on the standard stepped technique (Fig. 1).

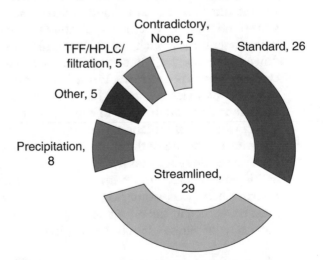

Fig. 1 Isolation technique survey of the EV-stem cell literature. Isolation methods were examined in 78 stem cell EV articles published through October of 2015, with inclusion and exclusion criteria as specified in the text. In 26 articles, a standard stepped centrifugation method was used for EV isolation ("standard"), including an intermediate step in the 5000–20,000 × g range. Twenty-eight described a "streamlined" protocol, omitting intermediate-speed steps. In some articles, these central "standard" or "streamlined" protocols were combined with additional purification steps, including gradients (not shown). Precipitation (usually PEG based) was used in eight articles. A combination of techniques, usually filtration, concentration, and gradients, is classified under "other" (5), while a combination of tangential flow filtration, high-performance liquid chromatography, and various filtration steps is denoted as "TFF/HPLC/filtration" (5). Five articles lacked clear methods or described what we interpreted as illogical isolation steps

Another 29 studies used a streamlined variant in which ultracentrifugation and wash steps followed a minimal $2000 \times g$ spin of 10 or 20 min. Since this protocol did not include intermediate spins (e.g., in the $10-20,000 \times g$ range), it is expected to yield a wider size range of more heterogeneous EVs than the standard protocol.

Two additional approaches bear mention. The first, reported in at least five publications on our list (Fig. 1), is a combination of techniques pioneered by the groups of Sai Kiang Lim and Dominique de Kleijn. This procedure initially concentrates cell culture conditioned supernatant by tangential flow filtration using a 100 kDa molecular weight cutoff (MWCO) filter. The concentrate is fractionated by high performance liquid chromatography, with the EV fraction further concentrated by MWCO filtration. Larger, presumably non-exosome EVs are removed by size filtration. This protocol appears to be a particularly useful method to process high-volume starting materials obtained from scaled-up cell culture.

The second alternative we would mention is precipitation, or "salting out," used in at least eight of our 78 publications (Fig. 1). Salting out employs a substance such as polyethylene glycol (PEG) to remove hydration shells from suspended bodies, allowing pelleting by low-speed centrifugation. Salting out is probably the least specific of the EV purification methods in current use [107], since all EV-sized bodies in a fluid are susceptible to salting out. Claims about precipitation as an EV- or exosome-specific method should be evaluated with caution. To be sure, PEG and other substances can be useful in reducing liquid volume prior to additional purification steps. If used this way, however, it may also be important to remove the polyethylene glycol or be vigilant for subsequent PEG-mediated effects.

Finally, a small but substantial group of articles did not report on their methods with sufficient detail to allow replication (Fig. 1). In some cases, the reported procedures if followed as described could not possibly have resulted in EV isolation. This unfortunate observation underlines the necessity of detailed methods as well as consultation, as appropriate, with subject matter experts on experimental procedures.

Ultracentrifugation, then, in one form or another, is the most widely used method to isolate EVs in stem cell-related studies. It should be noted that several recent reports have challenged aspects of the stepped centrifugation method. For example, ultracentrifugation, like most methods, will not yield completely pure vesicles [22], and some have reported that EVs may aggregate or lose a degree of functionality during ultracentrifugation [107, 108]. Size exclusion chromatography (SEC) has been advanced as a technique that results in fewer contaminants and subjects the EVs to less physical stress [109, 110]. These concerns should be considered carefully. However, the same concerns may not be relevant to all studies and are not shared by all. From a historical perspective, differential ultracentrifugation has been the mainstay of virology research for the better part of a century. While popularity and widespread use do not mean that any method is perfect, prudence is warranted and all implications of new approaches should be probed carefully before well-known techniques are abandoned completely. For the foreseeable future, ultracentrifugation will remain as one of the important workhorses in the EV isolation stable.

Since ultracentrifugation remains such an important technique in the field, we will now review the basic procedure. In Sect. 4, we will pay close attention to sources of variability in the methods, with particular emphasis on the importance of standardizing pellet resuspension. We also provide our protocol for a commonly used single particle tracking approach to characterize isolated particles, NanoSight nanoparticle tracking analysis (NTA). This method returns concentration and size profile for a population of particles such as EVs. There are many additional characterization methods, some of which are mentioned in Sect. 4.

2 Materials

2.1 Differential Centrifugation

Washed ultracentrifuge tubes. These tubes are often stored in cardboard boxes and may accumulate dust and debris. We recommend upside-down storage in sealed plastic storage boxes and rinsing well with distilled water or phosphate-buffered saline (PBS) before use to remove any debris.

Centrifuges appropriate for processing large volumes of cell culture conditioned medium, including an ultracentrifuge capable of at least 100,000 rcf. Although we use swinging bucket rotors exclusively, others use fixed-angle rotors and find that they are uniquely able to separate several distinct subpopulations of particles [111].

2.2 NanoSight Nanoparticle Tracking Analysis

NanoSight NS500 nanoparticle tracking instrument with syringe pump (Malvern). The instrument is equipped with one laser. Several laser wavelengths are available.

Computer with the latest version of nanoparticle tracking analysis software installed. Here, we refer to features in version 3.1.

Data storage system (at least several terabytes) if raw data files are to be retained. Raw data files can be revisited and reanalyzed at different settings; processed data cannot.

Filtered or otherwise highly purified distilled water and PBS or other diluent.

3 Methods

3.1 Differential Centrifugation of Cell Culture Conditioned Medium

See Fig. 2 for a flow chart of the procedure.

1. Collect CM (see **Notes 1.a** and **1.b**). Note that serum-free growth conditions should be used to ensure that EVs have been released from the cells of interest, not derived from serum

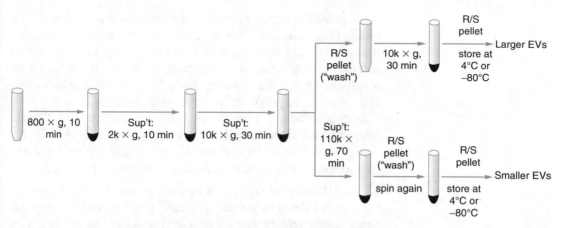

Fig. 2 Standard differential centrifugation procedure. Cells and debris are removed by lower speed spins (*left*), and then the supernatant is centrifuged at 10,000 ("10 k") × g. The pellet can be saved and (optionally) washed one or more times to obtain larger EVs (*top*), while the supernatant is centrifuged at 110 k × g for 70 min to enrich smaller EVs (*bottom*). Supernatant from the high-speed spin can be saved for analysis of protein complexes and other non-pelleted materials, while pellets are resuspended (R/S) in a buffer appropriate for downstream analyses. Isolated fractions can be stored at 4 °C for several days or at −80 °C longer-term

or other sources. Many stem cell media are defined and serum-free. When in doubt, ultracentrifuge or filter particle-containing media components to deplete them of exogenous EVs. A medium-only centrifugation is recommended as a control. It is also recommended to read a medium blank by NanoSight to ascertain background particle counts. Transfer CM to appropriately sized vessels for centrifugation. 50 mL conical tubes are often used for this step.

2. Spin CM at 300–800 × *g* for 10–15 min for removal of cells and large aggregates. It should not be necessary to dilute CM as one might, for example, plasma, saliva, and other relatively viscous fluids. We typically perform a spin of 800 × *g* for 10 min at room temperature. Pipette supernatant into new vessels, being careful not to disturb any pellet by leaving approximately 10% of the volume behind.

3. Spin transferred supernatant at 2000 × *g* for 10 min for removal of debris. Pipette supernatant into pre-rinsed ultracentrifuge tubes, again leaving some fluid behind to avoid disturbing the pelleted material.

4. Spin supernatant at 10,000 × *g* for 30 min to deplete remaining debris and large extracellular vesicles (see **Notes 1.c–1.e**). Transfer supernatant to new ultracentrifuge tubes as above. After draining remaining fluid, the pellet from this step, enriched in larger EVs including so-called microvesicles, may be retained and resuspended for one or more wash steps or resuspended in smaller volume for analysis and/or storage.

5. Spin supernatant at 110,000 × *g* for 70 min to enrich smaller EVs. Remove most supernatant by pipette and drain the remainder.

6. Resuspend the pellet. Note that the pellet will probably not be visible (see **Note 1.f**). The exact procedure for pellet resuspension is rarely discussed, and this may be an important source of variability within and between studies. Our protocol for pellet resuspension is: After draining the tube, use a micropipettor to introduce an appropriate volume of PBS (usually 400 μL to 1 mL). See **Note 1.g** on choice of resuspension buffer. Pipette up and down vigorously ten times. Vortex tube for 30 s. Place on ice for 20 min. Repeat the pipetting and vortexing steps, then transfer resuspended EVs to a new tube for storage.

7. Wash. Washing—that is, resuspending the pellet, filling the tube with PBS or other diluent, and spinning again—removes some contaminants but will also lead to some loss of EVs. The number of washes should be kept equal within an experiment or study, but beyond this consideration, a set number of washes cannot be prescribed. Whether or not a wash step is performed, and the extent of washing, must be determined by balancing time constraints, tolerance for contaminants, and yield.

8. After the desired number of washes, resuspend again, usually in a small volume. The resuspended pellet can be stored at 4 °C if it is to be analyzed or used within several days or at −80 °C for more extended periods.

It is essential to report on all variables necessary for replication of the procedure (*see* **Note 1.h**). Additional purification steps can be taken, and alternative methods for EV isolation are available.

3.2 Counting and Sizing by NanoSight Single Particle Tracking

See illustrations in Fig. 3.

Single particle tracking is a widely used method to obtain particle concentration and size profiles [112–114]. Particles in a viewing chamber are illuminated by laser, and scattered light is recorded by a high sensitivity camera through a microscope objective. A succession of video frames is analyzed by software to estimate (1) concentration and (2) size of each particle. Since particles are tracked through multiple frames, an average "jump" distance can be obtained for each particle. Plugging this value into the Stokes-Einstein equation results in a hydrodynamic sphere equivalent radius for each detected particle. In fluorescence mode, the instrument may detect fluor-bound particles [112]. Several particle tracking instruments are available; here, we focus on the NanoSight, offered by Malvern. Our system is accurate within a concentration range of approximately 1E8 to 1E9 particles per mL. A syringe pump allows smooth, continuous loading for best accuracy of concentration estimates; loading by peristaltic pump can also be done. See **Notes 2.a–2.e** for technical considerations related to day-to-day use of the instrument.

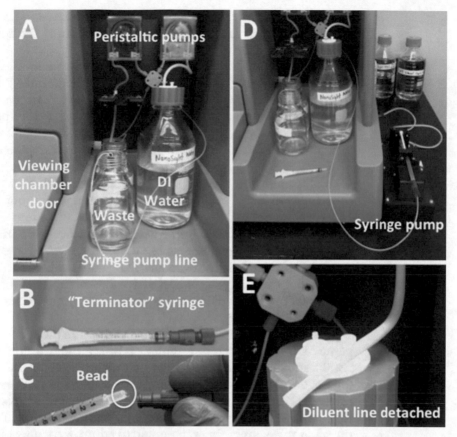

Fig. 3 NanoSight NS500 configuration for syringe pump usage. (**a**) Setup for fluidics priming. The line that connects to the syringe pump during syringe pump usage is routed instead to the waste. (**b**) A "terminator" syringe for the second step of priming is made by Parafilming the syringe barrel for immobility. However, any block will work. (**c**) Sample loaded into a syringe. The connector should be attached after "beading" a small amount of sample at the syringe tip. (**d**) Configuration of tubing for syringe pump usage. (**e**) To empty fluidics, the diluent line is removed from the diluent reservoir cap

3.2.1 Setup

With the instrument on, open the nanoparticle tracking analysis (NTA) software. The comments in this section refer to NTA version 3.1.

Prime the system: in the software interface, select the "Hardware" tab and "Prime Fluidics." Follow the software prompts to prime the system after confirming that the intake is drawing from a distilled water container and the tubing normally connected to the syringe pump routes instead to a waste container.

After completing one priming cycle, connect the syringe tubing either to a blocking plug or a "terminator syringe," consisting of a syringe barrel and plunger immobilized by, e.g., securing the plunger to the barrel with Parafilm.

Repeat the prime procedure to prime the viewing chamber.

If desired, replace water with PBS or other diluent by connecting PBS or other solution to the intake and repeating the last priming procedure.

Blank check: Close the NanoSight viewing door. Under the "Capture" tab, select "Start Camera." Press the "Scatter" button and make sure the stage is in the correct position. With the camera turned to maximum, no more than about three particles per field should be observed. A large number of particles would suggest that additional filtration or other purification steps are needed.

3.2.2   Loading Sample

Vortex sample. The sample should be diluted such that particle concentration is within the optimal reading range of about 1E8 to 1E9 particles per mL (see **Note 2.f** for recommendations on dilutions and dilution series). Immediately load at least 800 µL into a 1 mL Luer-lock syringe. Remember to flush the syringe line tubing if another sample has just been analyzed: add 1 mL of the diluent of choice to a syringe, connect it to the tubing, and manually (SLOWLY) push the plunger. Observe the results live on screen to assess proper purging.

Connect the tubing to the syringe and secure the syringe in the pump. Under the "syringe pump" tab, set the infusion rate to "800" and press "Infuse." Load approximately 400 µL. Switch camera on and evaluate the live image on screen for the presence of particles/change in number of particles. Stop the infusion when the sample has fully advanced into the viewing chamber. Position the stage at "Scatter" for each sample to ensure a comparable viewing window for all samples.

3.2.3   Video Recording

To run a script, open the "SOP" tab and double-click on the desired script. Script control is used for ease and for uniformity of recording; scripts can be customized easily. Once the script is opened, the steps will appear on screen for verification. For syringe pump readings, we recommend a script with at least three readings of 1 min each. If the peristaltic system is used, more reads of shorter duration may be appropriate. However, the appropriate numbers and lengths of reads must be determined for each research question. If the script does not include a prompt to set the infusion rate, set it prior to running the script. The infusion rate is typically between 20 and 50, depending on concentration (for higher concentrations, use a lower rate).

Select a camera setting appropriate for all samples to be read in a session. Alternatively the "Auto Setup" feature of the software can be used. This feature automatically selects settings based on the density and brightness of particles in the sample and is said to allow comparison of results obtained with different settings. However, we have not strictly verified this across a wide range of settings and recommend using the same settings for all samples in a group when possible.

Set the base file name and select "Run." See **Note 2.g** about minimizing vibration.

Upon script completion, flush the syringe line as in B, above.

3.2.4 Cleaning and Emptying Fluidics

After recording the final sample in a session and flushing, change the diluent source from PBS (or other diluent) to water by removing the top cap from the diluent container and attaching it to the water container.

Fill a syringe with distilled water and flush the syringe tubing manually three times; keep the camera on to make sure particle background drops to fewer than three particles per screen at camera level 16 (maximum setting).

Select "Empty Fluidics." Follow prompts on the screen as with the prime fluidics program. After the procedure completes, connect the terminator tubing to the syringe pump and repeat the process. The purpose of the emptying procedure is to remove most of the liquid from the lines to allow drying with limited residue deposition.

3.2.5 Data Processing

In the software interface, select the "Analysis" tab, then the "Process" tab. Select "Open Experiment" and select files that are to be analyzed together. (Note that analysis can be done for all samples together, or for each sample separately, as long as the same settings are used.)

Increase the screen gain to optimize particle viewing in the viewing window. (Note: this affects the on-screen appearance only and does not affect measured concentrations.)

Toggle the detection threshold to achieve a balance between "real" (represented by a red cross) and "spurious" (blue cross) detection. Roughly, one might expect 10–100 red crosses per frame (as indicated in the lower right corner of the analysis window) and fewer than 5 or so blue crosses. Scroll through the video to ensure that this ratio remains relatively consistent.

Highlight all files to be processed together and select "Process Selected Files." The videos will be processed in sequence.

Select "Export Results." Output will be one Excel file with bin data and one PDF with graphs and other average data. If the results are not exported, the Excel and PDF files will not be generated.

See **Note 2.h** on methods reporting and **Note 2.i** on orthogonal approaches to particle sizing and counting. **Note 2.j** emphasizes the necessity of additional assays for EV characterization.

4 Notes

1. Differential centrifugation.

   (a) For CM collection, the optimal period of culture must be determined empirically based on cell conditions and responses and kinetics of EV release. It is especially important to check cell viability, since cell death may result in different types of EVs in the medium [115].

(b) CM can be collected and stored prior to EV isolation, provided the initial, slower centrifugation steps (through at least $2000 \times g$) are done before storage to prevent artifactual formation of vesicles from cells and cellular debris. Clarified CM can then be stored at 4 °C for a week or more or frozen longer-term.

(c) For high-speed centrifugation steps, remember always to fill each tube to the top with an appropriate diluent (such as PBS) if CM does not fill the tube. The force of ultracentrifugation will crush tubes that are not full.

(d) Keep in mind that centrifugation is a physical process that does not purify any fraction absolutely. Small bodies at the very bottom of the column may pellet with slow spins, while large bodies at the top of the column may not pellet.

(e) After the initial, low-speed spins, centrifugation should be done with only gentle braking to avoid disturbing pellets. Consult with the manufacturer of your instrument if you have concerns about braking options and speeds.

(f) The ultracentrifuge pellet will probably not be visible. If using fixed-angle rotors, be sure to mark the side of the tube that will harbor the pellet. Attempt to standardize the pellet resuspension procedure as much as possible, as encouraged above, since resuspension is often difficult and can be a significant source of variability.

(g) PBS might not be the optimal resuspension buffer for all downstream applications. For example, if electron micrography will be performed, phosphates may not be compatible with negative stains. Here, another buffer, such as sodium cacodylate, could be used.

(h) When reporting results, be sure to include in the methods the exact centrifuges, rotors, and tubes used, along with relative centrifugal force, time, temperature, and brake settings. It is also helpful to provide the k factor, or clearance factor, for the rotor/tube combination used, although this can be derived from the above information.

2. NanoSight nanoparticle tracking analysis.

(a) Consult with NanoSight training videos or technical assistance to learn about stage positioning, which we do not address here.

(b) The NanoSight instrument can be left on with the stage under the camera, since the laser is automatically turned off when not in use. It is recommended to turn off the machine periodically, especially if it is not expected to be used for several days.

(c) The top gasket that guides sample into the viewing chamber has a seal that may degrade or loosen over time. Repeated removal and reapplication shortens the working life of this seal; however, removal is often necessary to eliminate recalcitrant air bubbles or to clean residue.

(d) Introduction of air into the fluidics is a common problem. Air in or near the viewing chamber is easily identified when particles move "diagonally" across the viewing screen instead of horizontally.

(e) Residues can be left behind during normal operation. These include salts from PBS and other diluents, precipitation reagents that may be used to enrich EVs, and detergents used to assess differential sensitivity. To minimize residue accumulation, the system is flushed well with water before shutdown. Consult with the manufacturer about substances and maximum concentrations that are compatible with the tubing, connections, and viewing chamber of your instrument.

(f) A high-concentration sample can sometimes appear to read within the recommended range of 1E8 to 1E9 particles per mL despite actually being more concentrated. For this reason, it may be useful, albeit time-consuming, to perform one or more dilutions (e.g., 1:10) of samples to identify those that do not drop in recorded concentration by the applied dilution factor and are thus out of range, requiring greater dilution.

(g) The NTA software can factor out some vibrations; however, when making recordings with the NanoSight, it is recommended to minimize sources of vibration, including nearby electronic or mechanical equipment and operator actions such as leaning on the table or typing on a computer keyboard. An anti-vibration table or pad for the instrument may also help to minimize vibrations.

(h) All relevant details of the recording and analysis methods should be reported in submitted manuscripts. These include dilution experiments, hardware and software specifications and versions, camera settings and automatic options, length of recording and number of recordings, scripts, analysis settings, and details of statistical analysis.

(i) Other approaches to concentration estimation and sizing include electron microscopy, certain forms of flow cytometry, dynamic light scattering [116, 117], and tunable resistive pulse sensing [118–120].

(j) NTA readings should not be misinterpreted as EV specific. Any particle of sufficient size and density can scatter light. As a result, NTA must be used alongside other techniques for EV characterization. We strongly recommend that researchers follow the ISEV "MISEV" minimal requirements when characterizing EVs [10], confirming the presence of EV-associated markers and the relative exclusion of cell-specific or cell-enriched markers by Western, ELISA, flow cytometry, immunogold EM, etc. It is possible that some characterization can be done by NanoSight using fluorescence mode to detect brightly conjugated affinity molecules [112]. However, only one channel can be assessed per sample, and bright signal is needed. Because of low sensitivity, we have, for example, been unsuccessful in using NTA with fluorescent lipid dyes to identify the fraction of our isolated particle populations that has a lipid membrane. Above all, other characterization methods must be used to complement NTA.

Note added in proof: The authors would like to refer to two recent articles that also address methods and standardization in the EV field [121, 122].

References

1. Yáñez-Mó M, Siljander PR-M, Andreu Z et al (2015) Biological properties of extracellular vesicles and their physiological functions. J Extracell Vesicles 4:27066

2. Colombo M, Raposo G, Théry C (2014) Biogenesis, secretion, and intercellular interactions of exosomes and other extracellular vesicles. Annu Rev Cell Dev Biol 30:255–289. doi:10.1146/annurev-cellbio-101512-122326

3. Di Vizio D, Morello M, Dudley AC et al (2012) Large oncosomes in human prostate cancer tissues and in the circulation of mice with metastatic disease. Am J Pathol 181:1573–1584. doi:10.1016/j.ajpath.2012.07.030

4. Chargaff E, West R (1946) The biological significance of the thromboplastic protein of blood. J Biol Chem 166:189–197

5. Wolf P (1967) The nature and significance of platelet products in human plasma. Br J Haematol 13:269–288

6. Anderson HC (1969) Vesicles associated with calcification in the matrix of epiphyseal cartilage. J Cell Biol 41:59–72

7. Bonucci E (1967) Fine structure of early cartilage calcification. J Ultrastruct Res 20:33–50

8. Anderson HC (1967) Electron microscopic studies of induced cartilage development and calcification. J Cell Biol 35:81–101

9. Kim D-K, Lee J, Kim SR et al (2015) EVpedia: a community web portal for extracellular vesicles research. Bioinformatics 31:933–939. doi:10.1093/bioinformatics/btu741

10. Lotvall J, Hill AF, Hochberg F et al (2014) Minimal experimental requirements for definition of extracellular vesicles and their functions: a position statement from the International Society for Extracellular Vesicles. J Extracell Vesicles 3:26913. doi:10.3402/jev.v3.26913

11. Mack M, Kleinschmidt A, Bruhl H et al (2000) Transfer of the chemokine receptor CCR5 between cells by membrane-derived microparticles: a mechanism for cellular human immunodeficiency virus 1 infection. Nat Med 6:769–775. doi:10.1038/77498

12. Lotvall J, Valadi H (2007) Cell to cell signalling via exosomes through esRNA. Cell Adh Migr 1:156–158

13. Ridder K, Sevko A, Heide J et al (2015) Extracellular vesicle-mediated transfer of functional RNA in the tumor microenvironment. Oncoimmunology 4:e1008371. doi:10.1080/2162402X.2015.1008371

14. Zomer A, Maynard C, Verweij FJ et al (2015) In vivo imaging reveals extracellular vesicle-mediated phenocopying of metastatic behavior. Cell 161:1046–1057. doi:10.1016/j.cell.2015.04.042

15. Iraci N, Leonardi T, Gessler F et al (2016) Focus on extracellular vesicles: physiological role and signalling properties of extracellular membrane vesicles. Int J Mol Sci 17:171. doi:10.3390/ijms17020171

16. Ciardiello C, Cavallini L, Spinelli C et al (2016) Focus on extracellular vesicles: new frontiers of cell-to-cell communication in cancer. Int J Mol Sci 17:175. doi:10.3390/ijms17020175

17. Gnecchi M, He H, Liang OD et al (2005) Paracrine action accounts for marked protection of ischemic heart by Akt-modified mesenchymal stem cells. Nat Med 11:367–368. doi:10.1038/nm0405-367

18. Lai RC, Arslan F, Lee MM et al (2010) Exosome secreted by MSC reduces myocardial ischemia/reperfusion injury. Stem Cell Res 4:214–222. doi:10.1016/j.scr.2009.12.003

19. Bruno S, Grange C, Deregibus MC et al (2009) Mesenchymal stem cell-derived microvesicles protect against acute tubular injury. J Am Soc Nephrol 20:1053–1067. doi:10.1681/ASN.2008070798

20. Lener T, Gimona M, Aigner L et al (2015) Applying extracellular vesicles based therapeutics in clinical trials--an ISEV position paper. J Extracell Vesicles 4:30087

21. Fischer UM, Harting MT, Jimenez F et al (2009) Pulmonary passage is a major obstacle for intravenous stem cell delivery: the pulmonary first-pass effect. Stem Cells Dev 18:683–692. doi:10.1089/scd.2008.0253

22. Webber J, Clayton A (2013) How pure are your vesicles? J Extracell Vesicles 2. doi:10.3402/jev.v2i0.19861. [pii]

23. Jo W, Jeong D, Kim J et al (2014) Microfluidic fabrication of cell-derived nanovesicles as endogenous RNA carriers. Lab Chip 14:1261–1269. doi:10.1039/c3lc50993a

24. Beall J (2016) Scholarly Open Access ("Beall's List") https://scholarlyoa.com/publishers/.

25. Phinney DG, Di Giuseppe M, Njah J et al (2015) Mesenchymal stem cells use extracellular vesicles to outsource mitophagy and shuttle microRNAs. Nat Commun 6:8472. doi:10.1038/ncomms9472

26. Wang Y, Fu B, Sun X et al (2015) Differentially expressed microRNAs in bone marrow mesenchymal stem cell-derived microvesicles in young and older rats and their effect on tumor growth factor-beta1-mediated epithelial-mesenchymal transition in HK2 cells. Stem Cell Res Ther 6:185. doi:10.1186/s13287-015-0179-x

27. Liu S, Liu D, Chen C et al (2015) MSC transplantation improves osteopenia via epigenetic regulation of notch signaling in lupus. Cell Metab 22:606–618. doi:10.1016/j.cmet.2015.08.018

28. Vandergriff AC, de Andrade JBM, Tang J et al (2015) Intravenous cardiac stem cell-derived exosomes ameliorate cardiac dysfunction in doxorubicin induced dilated cardiomyopathy. Stem Cells Int 2015:960926. doi:10.1155/2015/960926

29. Wang X, Gu H, Qin D et al (2015) Exosomal miR-223 contributes to mesenchymal stem cell-elicited cardioprotection in polymicrobial sepsis. Sci Rep 5:13721. doi:10.1038/srep13721

30. Ko S-F, Yip H-K, Zhen Y-Y et al (2015) Adipose-derived mesenchymal stem cell exosomes suppress hepatocellular carcinoma growth in a rat model: apparent diffusion coefficient, natural killer T-cell responses, and histopathological features. Stem Cells Int 2015:853506. doi:10.1155/2015/853506

31. Conigliaro A, Costa V, Lo Dico A et al (2015) CD90+ liver cancer cells modulate endothelial cell phenotype through the release of exosomes containing H19 lncRNA. Mol Cancer 14:155. doi:10.1186/s12943-015-0426-x

32. Hu G, Li Q, Niu X et al (2015) Exosomes secreted by human-induced pluripotent stem cell-derived mesenchymal stem cells attenuate limb ischemia by promoting angiogenesis in mice. Stem Cell Res Ther 6:10. doi:10.1186/scrt546

33. Zhao Y, Sun X, Cao W et al (2015) Exosomes derived from human umbilical cord mesenchymal stem cells relieve acute myocardial ischemic injury. Stem Cells Int 2015:761643. doi:10.1155/2015/761643

34. Monsel A, Zhu Y, Gennai S et al (2015) Therapeutic effects of human mesenchymal stem cell-derived microvesicles in severe pneumonia in mice. Am J Respir Crit Care Med 192:324–336. doi:10.1164/rccm.201410-1765OC

35. Bobis-Wozowicz S, Kmiotek K, Sekula M et al (2015) Human induced pluripotent stem cell-derived microvesicles transmit RNAs and proteins to recipient mature heart cells modulating cell fate and behavior. Stem Cells 33:2748–2761. doi:10.1002/stem.2078

36. Jarmalaviciute A, Tunaitis V, Pivoraite U et al (2015) Exosomes from dental pulp stem cells rescue human dopaminergic neurons from 6-hydroxy-dopamine-induced apoptosis. Cytotherapy 17:932–939. doi:10.1016/j.jcyt.2014.07.013

37. Khan M, Nickoloff E, Abramova T et al (2015) Embryonic stem cell-derived exosomes promote endogenous repair mechanisms and enhance cardiac function following myocardial infarction. Circ Res 117:52–64. doi:10.1161/CIRCRESAHA.117.305990

38. Pivoraite U, Jarmalaviciute A, Tunaitis V et al (2015) Exosomes from human dental pulp stem cells suppress carrageenan-induced acute inflammation in mice. Inflammation 38:1933–1941. doi:10.1007/s10753-015-0173-6

39. Shabbir A, Cox A, Rodriguez-Menocal L et al (2015) Mesenchymal stem cell exosomes induce proliferation and migration of normal and chronic wound fibroblasts, and enhance angiogenesis in vitro. Stem Cells Dev 24:1635–1647. doi:10.1089/scd.2014.0316

40. Nakamura Y, Miyaki S, Ishitobi H et al (2015) Mesenchymal-stem-cell-derived exosomes accelerate skeletal muscle regeneration. FEBS Lett 589:1257–1265. doi:10.1016/j.febslet.2015.03.031

41. Yang J, Gao F, Zhang Y et al (2015) Buyang Huanwu decoction (BYHWD) enhances angiogenic effect of mesenchymal stem cell by upregulating VEGF expression after focal cerebral ischemia. J Mol Neurosci 56:898–906. doi:10.1007/s12031-015-0539-0

42. Ju G, Cheng J, Zhong L et al (2015) Microvesicles derived from human umbilical cord mesenchymal stem cells facilitate tubular epithelial cell dedifferentiation and growth via hepatocyte growth factor induction. PLoS One 10:e0121534. doi:10.1371/journal.pone.0121534

43. Raimondo S, Saieva L, Corrado C et al (2015) Chronic myeloid leukemia-derived exosomes promote tumor growth through an autocrine mechanism. Cell Commun Signal 13:8. doi:10.1186/s12964-015-0086-x

44. Zhang J, Guan J, Niu X et al (2015) Exosomes released from human induced pluripotent stem cells-derived MSCs facilitate cutaneous wound healing by promoting collagen synthesis and angiogenesis. J Transl Med 13:49. doi:10.1186/s12967-015-0417-0

45. Vyas N, Walvekar A, Tate D et al (2014) Vertebrate hedgehog is secreted on two types of extracellular vesicles with different signaling properties. Sci Rep 4:7357. doi:10.1038/srep07357

46. Maredziak M, Marycz K, Lewandowski D et al (2015) Static magnetic field enhances synthesis and secretion of membrane-derived microvesicles (MVs) rich in VEGF and BMP-2 in equine adipose-derived stromal cells (EqASCs)-a new approach in veterinary regenerative medicine. In Vitro Cell Dev Biol Anim 51:230–240. doi:10.1007/s11626-014-9828-0

47. Blazquez R, Sanchez-Margallo FM, de la Rosa O et al (2014) Immunomodulatory potential of human adipose mesenchymal stem cells derived exosomes on in vitro stimulated T cells. Front Immunol 5:556. doi:10.3389/fimmu.2014.00556

48. Garcia-Contreras M, Vera-Donoso CD, Hernandez-Andreu JM et al (2014) Therapeutic potential of human adipose-derived stem cells (ADSCs) from cancer patients: a pilot study. PLoS One 9:e113288. doi:10.1371/journal.pone.0113288

49. Sims B, Gu L, Krendelchtchikov A, Matthews QL (2014) Neural stem cell-derived exosomes mediate viral entry. Int J Nanomedicine 9:4893–4897

50. Gray WD, French KM, Ghosh-Choudhary S et al (2015) Identification of therapeutic covariant microRNA clusters in hypoxia-treated cardiac progenitor cell exosomes using systems biology. Circ Res 116:255–263. doi:10.1161/CIRCRESAHA.116.304360

51. Pascucci L, Alessandri G, Dall'Aglio C et al (2014) Membrane vesicles mediate pro-angiogenic activity of equine adipose-derived mesenchymal stromal cells. Vet J 202:361–366. doi:10.1016/j.tvjl.2014.08.021

52. Jeong D, Jo W, Yoon J et al (2014) Nanovesicles engineered from ES cells for enhanced cell proliferation. Biomaterials 35:9302–9310. doi:10.1016/j.biomaterials.2014.07.047

53. Chen J, An R, Liu Z et al (2014) Therapeutic effects of mesenchymal stem cell-derived microvesicles on pulmonary arterial hypertension in rats. Acta Pharmacol Sin 35:1121–1128. doi:10.1038/aps.2014.61

54. Ono M, Kosaka N, Tominaga N et al (2014) Exosomes from bone marrow mesenchymal stem cells contain a microRNA that promotes dormancy in metastatic breast cancer cells. Sci Signal 7:–ra63. doi:10.1126/scisignal.2005231

55. Zhang B, Wang M, Gong A et al (2015) HucMSC-exosome mediated-Wnt4 signaling is required for cutaneous wound healing. Stem Cells 33:2158–2168. doi:10.1002/stem.1771

56. Tan CY, Lai RC, Wong W et al (2014) Mesenchymal stem cell-derived exosomes promote hepatic regeneration in drug-induced liver injury models. Stem Cell Res Ther 5:76. doi:10.1186/scrt465

57. Favaro E, Carpanetto A, Lamorte S et al (2014) Human mesenchymal stem cell-derived microvesicles modulate T cell response to islet antigen glutamic acid decarboxylase in patients with type 1 diabetes. Diabetologia 57:1664–1673. doi:10.1007/s00125-014-3262-4

58. Lin S-S, Zhu B, Guo Z-K et al (2014) Bone marrow mesenchymal stem cell-derived microvesicles

protect rat pheochromocytoma PC12 cells from glutamate-induced injury via a PI3K/Akt dependent pathway. Neurochem Res 39:922–931. doi:10.1007/s11064-014-1288-0

59. Hazawa M, Tomiyama K, Saotome-Nakamura A et al (2014) Radiation increases the cellular uptake of exosomes through CD29/CD81 complex formation. Biochem Biophys Res Commun 446:1165–1171. doi:10.1016/j.bbrc.2014.03.067

60. Raisi A, Azizi S, Delirezh N et al (2014) The mesenchymal stem cell-derived microvesicles enhance sciatic nerve regeneration in rat: a novel approach in peripheral nerve cell therapy. J Trauma Acute Care Surg 76:991–997. doi:10.1097/TA.0000000000000186

61. Zou X, Zhang G, Cheng Z et al (2014) Microvesicles derived from human Wharton's Jelly mesenchymal stromal cells ameliorate renal ischemia-reperfusion injury in rats by suppressing CX3CL1. Stem Cell Res Ther 5:40. doi:10.1186/scrt428

62. Papi A, De Carolis S, Bertoni S et al (2014) PPARgamma and RXR ligands disrupt the inflammatory cross-talk in the hypoxic breast cancer stem cells niche. J Cell Physiol 229:1595–1606. doi:10.1002/jcp.24601

63. Choi M, Ban T, Rhim T (2014) Therapeutic use of stem cell transplantation for cell replacement or cytoprotective effect of microvesicle released from mesenchymal stem cell. Mol Cells 37:133–139. doi:10.14348/molcells.2014.2317

64. Bourkoula E, Mangoni D, Ius T et al (2014) Glioma-associated stem cells: a novel class of tumor-supporting cells able to predict prognosis of human low-grade gliomas. Stem Cells 32:1239–1253. doi:10.1002/stem.1605

65. Tan SS, Yin Y, Lee T et al (2013) Therapeutic MSC exosomes are derived from lipid raft microdomains in the plasma membrane. J Extracell Vesicles 2. doi:10.3402/jev.v2i0.22614

66. Li Q, Eades G, Yao Y et al (2014) Characterization of a stem-like subpopulation in basal-like ductal carcinoma in situ (DCIS) lesions. J Biol Chem 289:1303–1312. doi:10.1074/jbc.M113.502278

67. Rahman MJ, Regn D, Bashratyan R, Dai YD (2014) Exosomes released by islet-derived mesenchymal stem cells trigger autoimmune responses in NOD mice. Diabetes 63:1008–1020. doi:10.2337/db13-0859

68. Feng Y, Huang W, Meng W et al (2014) Heat shock improves Sca-1+ stem cell survival and directs ischemic cardiomyocytes toward a prosurvival phenotype via exosomal transfer: a critical role for HSF1/miR-34a/HSP70 pathway. Stem Cells 32:462–472. doi:10.1002/stem.1571

69. Munoz JL, Bliss SA, Greco SJ et al (2013) Delivery of functional anti-miR-9 by mesenchymal stem cell-derived exosomes to glioblastoma multiforme cells conferred chemosensitivity. Mol Ther Nucleic Acids 2:e126. doi:10.1038/mtna.2013.60

70. Yu B, Gong M, Wang Y et al (2013) Cardiomyocyte protection by GATA-4 gene engineered mesenchymal stem cells is partially mediated by translocation of miR-221 in microvesicles. PLoS One 8:e73304. doi:10.1371/journal.pone.0073304

71. Mokarizadeh A, Rezvanfar M-A, Dorostkar K, Abdollahi M (2013) Mesenchymal stem cell derived microvesicles: trophic shuttles for enhancement of sperm quality parameters. Reprod Toxicol 42:78–84. doi:10.1016/j.reprotox.2013.07.024

72. Zhu Y-G, Feng X-M, Abbott J et al (2014) Human mesenchymal stem cell microvesicles for treatment of *Escherichia coli* endotoxin-induced acute lung injury in mice. Stem Cells 32:116–125. doi:10.1002/stem.1504

73. Coulson-Thomas VJ, Caterson B, Kao WW-Y (2013) Transplantation of human umbilical mesenchymal stem cells cures the corneal defects of mucopolysaccharidosis VII mice. Stem Cells 31:2116–2126. doi:10.1002/stem.1481

74. Lin R, Wang S, Zhao RC (2013) Exosomes from human adipose-derived mesenchymal stem cells promote migration through Wnt signaling pathway in a breast cancer cell model. Mol Cell Biochem 383:13–20. doi:10.1007/s11010-013-1746-z

75. Xin H, Li Y, Liu Z et al (2013) MiR-133b promotes neural plasticity and functional recovery after treatment of stroke with multipotent mesenchymal stromal cells in rats via transfer of exosome-enriched extracellular particles. Stem Cells 31:2737–2746. doi:10.1002/stem.1409

76. Zhou Y, Xu H, Xu W et al (2013) Exosomes released by human umbilical cord mesenchymal stem cells protect against cisplatin-induced renal oxidative stress and apoptosis in vivo and in vitro. Stem Cell Res Ther 4:34. doi:10.1186/scrt194

77. Roccaro AM, Sacco A, Maiso P et al (2013) BM mesenchymal stromal cell-derived exosomes facilitate multiple myeloma progression. J Clin Invest 123:1542–1555. doi:10.1172/JCI66517

78. Askar SFA, Ramkisoensing AA, Atsma DE et al (2013) Engraftment patterns of human adult mesenchymal stem cells expose electrotonic and paracrine proarrhythmic mechanisms in myocardial cell cultures. Circ Arrhythm Electrophysiol 6:380–391. doi:10.1161/CIRCEP.111.000215

79. Arslan F, Lai RC, Smeets MB et al (2013) Mesenchymal stem cell-derived exosomes increase ATP levels, decrease oxidative stress and activate PI3K/Akt pathway to enhance myocardial viability and prevent adverse remodeling after myocardial ischemia/reperfusion injury. Stem Cell Res 10:301–312. doi:10.1016/j.scr.2013.01.002

80. Katsman D, Stackpole EJ, Domin DR, Farber DB (2012) Embryonic stem cell-derived microvesicles induce gene expression changes in Muller cells of the retina. PLoS One 7:e50417. doi:10.1371/journal.pone.0050417

81. Lee C, Mitsialis SA, Aslam M et al (2012) Exosomes mediate the cytoprotective action of mesenchymal stromal cells on hypoxia-induced pulmonary hypertension. Circulation 126:2601–2611.   doi:10.1161/CIRCULATIONAHA.112.114173

82. Patel AN, Vargas V, Revello P, Bull DA (2013) Mesenchymal stem cell population isolated from the subepithelial layer of umbilical cord tissue. Cell Transplant 22:513–519. doi:10.3727/096368912X655064

83. Ratajczak J, Kucia M, Mierzejewska K et al (2013) Paracrine proangiopoietic effects of human umbilical cord blood-derived purified CD133+ cells–implications for stem cell therapies in regenerative medicine. Stem Cells Dev   22:422–430.   doi:10.1089/scd.2012.0268

84. Reis LA, Borges FT, Simoes MJ et al (2012) Bone marrow-derived mesenchymal stem cells repaired but did not prevent gentamicin-induced acute kidney injury through paracrine effects in rats. PLoS One 7:e44092. doi:10.1371/journal.pone.0044092

85. Iglesias DM, El-Kares R, Taranta A et al (2012) Stem cell microvesicles transfer cystinosin to human cystinotic cells and reduce cystine accumulation in vitro. PLoS One 7:e42840. doi:10.1371/journal.pone.0042840

86. Lai RC, Tan SS, Teh BJ et al (2012) Proteolytic potential of the MSC exosome proteome: implications for an exosome-mediated delivery of therapeutic proteasome. Int J Proteomics 2012:971907. doi:10.1155/2012/971907

87. Fonsato V, Collino F, Herrera MB et al (2012) Human liver stem cell-derived microvesicles inhibit hepatoma growth in SCID mice by delivering antitumor microRNAs.   Stem   Cells   30:1985–1998. doi:10.1002/stem.1161

88. van Koppen A, Joles JA, van Balkom BWM et al (2012) Human embryonic mesenchymal stem cell-derived conditioned medium rescues kidney function in rats with established chronic kidney disease. PLoS One 7:e38746. doi:10.1371/journal.pone.0038746

89. Mackie AR, Klyachko E, Thorne T et al (2012) Sonic hedgehog-modified human CD34+ cells preserve cardiac function after acute myocardial infarction.   Circ   Res   111:312–321. doi:10.1161/CIRCRESAHA.112.266015

90. Cantaluppi V, Gatti S, Medica D et al (2012) Microvesicles derived from endothelial progenitor cells protect the kidney from ischemia-reperfusion   injury   by microRNA-dependent reprogramming of resident renal cells. Kidney Int 82:412–427. doi:10.1038/ki.2012.105

91. Strassburg S, Hodson NW, Hill PI et al (2012) Bi-directional exchange of membrane components occurs during co-culture of mesenchymal stem cells and nucleus pulposus cells. PLoS One 7:e33739. doi:10.1371/journal.pone.0033739

92. He J, Wang Y, Sun S et al (2012) Bone marrow stem cells-derived microvesicles protect against renal injury in the mouse remnant kidney model. Nephrology (Carlton) 17:493–500.doi:10.1111/j.1440-1797.2012.01589.x

93. Sahoo S, Klychko E, Thorne T et al (2011) Exosomes from human CD34(+) stem cells mediate their proangiogenic paracrine activity. Circ Res 109:724–728. doi:10.1161/CIRCRESAHA.111.253286. [pii]

94. Grange C, Tapparo M, Collino F et al (2011) Microvesicles released from human renal cancer stem cells stimulate angiogenesis and formation of lung premetastatic niche. Cancer Res   71:5346–5356.   doi:10.1158/0008-5472.CAN-11-0241

95. Bauer N, Wilsch-Brauninger M, Karbanova J et al (2011) Haematopoietic stem cell differentiation promotes the release of prominin-1/CD133-containing   membrane vesicles--a role of the endocytic-exocytic pathway. EMBO Mol Med 3:398–409. doi:10.1002/emmm.201100147

96. Gatti S, Bruno S, Deregibus MC et al (2011) Microvesicles derived from human adult mesenchymal stem cells protect against ischaemia-reperfusion-induced acute and chronic kidney injury. Nephrol Dial Transplant 26:1474–1483. doi:10.1093/ndt/gfr015

97. Herrera MB, Fonsato V, Gatti S et al (2010) Human liver stem cell-derived microvesicles accelerate hepatic regeneration in hepatectomized rats. J Cell Mol Med 14:1605–1618. doi:10.1111/j.1582-4934.2009.00860.x

98. Yuan A, Farber EL, Rapoport AL et al (2009) Transfer of microRNAs by embryonic stem cell microvesicles. PLoS One 4:e4722. doi:10.1371/journal.pone.0004722

99. Ratajczak J, Miekus K, Kucia M et al (2006) Embryonic stem cell-derived microvesicles reprogram hematopoietic progenitors: evidence

for horizontal transfer of mRNA and protein delivery. Leukemia 20:847–856. doi:10.1038/sj.leu.2404132

100. Zhou J, Ghoroghi S, Benito-Martin A et al (2016) Characterization of induced pluripotent stem cell microvesicle genesis, morphology and pluripotent content. Sci Rep 6:19743. doi:10.1038/srep19743

101. Gould SJ, Raposo G (2013) As we wait: coping with an imperfect nomenclature for extracellular vesicles. J Extracell Vesicles 2. doi:10.3402/jev.v2i0.20389. [pii]

102. Thery C, Amigorena S, Raposo G, Clayton A (2006) Isolation and characterization of exosomes from cell culture supernatants and biological fluids. Curr Protoc Cell Biol Chapter 3:Unit 3.22. doi: 10.1002/0471143030.cb0322s30

103. Konadu KA, Huang MB, Roth W, et al (2016) Isolation of Exosomes from the Plasma of HIV-1 Positive Individuals. J Vis Exp 107: e53495. doi: 10.3791/53495

104. Raposo G, Nijman HW, Stoorvogel W et al (1996) B lymphocytes secrete antigen-presenting vesicles. J Exp Med 183:1161–1172

105. Palma J, Yaddanapudi SC, Pigati L et al (2012) MicroRNAs are exported from malignant cells in customized particles. Nucleic Acids Res 40:9125–9138. doi:10.1093/nar/gks656. [pii]

106. Aalberts M, van Dissel-Emiliani FM, van Adrichem NP et al (2012) Identification of distinct populations of prostasomes that differentially express prostate stem cell antigen, annexin A1, and GLIPR2 in humans. Biol Reprod 86:82. doi:10.1095/biolreprod.111.095760. [pii]

107. Lobb RJ, Becker M, Wen SW et al (2015) Optimized exosome isolation protocol for cell culture supernatant and human plasma. J Extracell Vesicles 4:27031

108. Linares R, Tan S, Gounou C et al (2015) High-speed centrifugation induces aggregation of extracellular vesicles. J Extracell Vesicles 4:29509

109. Boing AN, van der Pol E, Grootemaat AE et al (2014) Single-step isolation of extracellular vesicles by size-exclusion chromatography. J Extracell Vesicles 3. doi:10.3402/jev.v3.23430. [pii]

110. Welton JL, Webber JP, Botos L-A et al (2015) Ready-made chromatography columns for extracellular vesicle isolation from plasma. J Extracell Vesicles 4:27269

111. Cvjetkovic A, Lotvall J, Lasser C (2014) The influence of rotor type and centrifugation time on the yield and purity of extracellular vesicles. J Extracell Vesicles 3. doi:10.3402/jev.v3.23111. [pii]

112. Dragovic RA, Gardiner C, Brooks AS et al (2011) Sizing and phenotyping of cellular vesicles using Nanoparticle Tracking Analysis. Nanomedicine 7:780–788. doi:10.1016/j.nano.2011.04.003. S1549–9634(11)00163–8 [pii]

113. Gardiner C, Ferreira YJ, Dragovic RA et al (2013) Extracellular vesicle sizing and enumeration by nanoparticle tracking analysis. J Extracell Vesicles 2:19671

114. van der Pol E, Coumans FAW, Grootemaat AE et al (2014) Particle size distribution of exosomes and microvesicles determined by transmission electron microscopy, flow cytometry, nanoparticle tracking analysis, and resistive pulse sensing. J Thromb Haemost 12:1182–1192. doi:10.1111/jth.12602

115. Witwer KW, Buzas EI, Bemis LT et al (2013) Standardization of sample collection, isolation and analysis methods in extracellular vesicle research. J Extracell Vesicles 2. doi:10.3402/jev.v2i0.20360. [pii]

116. Varga Z, Yuana Y, Grootemaat AE et al (2014) Towards traceable size determination of extracellular vesicles. J Extracell Vesicles 3. doi:10.3402/jev.v3.23298

117. Tosar JP, Gambaro F, Sanguinetti J et al (2015) Assessment of small RNA sorting into different extracellular fractions revealed by high-throughput sequencing of breast cell lines. Nucleic Acids Res 43:5601–5616. doi:10.1093/nar/gkv432

118. Kozak D, Anderson W, Vogel R, Trau M (2011) Advances in resistive pulse sensors: devices bridging the void between molecular and microscopic detection. Nano Today 6:531–545. doi:10.1016/j.nantod.2011.08.012

119. Coumans FAW, van der Pol E, Böing AN et al (2014) Reproducible extracellular vesicle size and concentration determination with tunable resistive pulse sensing. J Extracell Vesicles 3:25922

120. Maas SLN, De Vrij J, Broekman MLD (2014) Quantification and size-profiling of extracellular vesicles using tunable resistive pulse sensing. J Vis Exp 92: e51623. doi: 10.3791/51623

121. Gardiner C, Di Vizio D, Sahoo S et al (2016) Techniques used for the isolation and characterization of extracellular vesicles: results of a worldwide survey. J Extracell Vesicles 5:32945. doi: 10.3402/jev.v5.32945.

122. EV-TRACK Consortium, Van Deun J, Mestdagh P et al (2017) EV-TRACK: transparent reporting and centralizing knowledge in extracellular vesicle research. Nat Methods 14(3):228-232. doi: 10.1038/nmeth.4185.

Chapter 15

Essential Requirements for Setting Up a Stem Cell Laboratory

Philip H. Schwartz and Robin L. Wesselschmidt

Abstract

This chapter provides an overview and general considerations for establishing a stem cell research laboratory. It provides practical information about establishing stem cell research laboratory operations with an emphasis on quality control/quality assurance as well as an overview and considerations regarding equipment and laboratory design. Stem cells require specialized cell culture expertise and a robust quality control system in order to maintain the identity of the stem cells within specified parameters or to develop robust processes for their expansion and differentiation. This chapter aims to provide a starting point for the establishment of a stem cell research laboratory with equal emphasis on the facility and operations.

Key words Laboratory design, Stem cells, Cell culture, Quality control, Standard operating procedures

1 Introduction

This chapter provides an overview of the critical issues that need to be addressed when establishing a stem cell laboratory in a research and development environment. The focus of the chapter is equally divided on the physical space and equipment as well as the establishing operations that lay the foundation for successful experimentation and reproducible results. The physical space requirements for the research/discovery stem cell laboratory are essentially the same as that required for other mammalian cell culture laboratories; it is the quality control and oversight that generally set the stem cell laboratory apart [1–3]. Stem cells require specialized cell culture expertise and a robust quality control system. Additionally, because these cells are derived or isolated from human tissues, there are generally more reporting requirements and regulatory oversight.

This chapter is divided into two main sections: facility overview, which addresses the physical space including infrastructure and laboratory design and equipment, and operations overview,

Amit K. Srivastava et al. (eds.), *Stem Cell Technologies in Neuroscience*, Neuromethods, vol. 126,
DOI 10.1007/978-1-4939-7024-7_15, © Springer Science+Business Media LLC 2017

which addresses the operation of a state-of-the-art stem cell laboratory with suggested cell banking strategy and quality assurance/control measures.

2   Facility Overview

Whether building a new laboratory or renovating existing laboratory space, facility design and upgrade issues will most likely be undertaken by a team comprised of people with the specialized knowledge and expertise that is required to address design and safety guidelines, health and safety regulations, and codes and standards for laboratory construction. At a minimum, this team usually includes the laboratory head/principal investigator, institutional facility manager, and an architect/engineer. During the design planning phase, the team aims to design a laboratory, including ancillary support space, that provides maximal support to personnel in order to safely and effectively reach the program goals within budget constraints. Below is a list of some of the key considerations in the infrastructure and design of a stem cell laboratory. This list is by no means exhaustive and is meant to provide the laboratory head/principal investigator and their staff an overview of the major considerations as the stem cell laboratory is planned with the help of the engineer and architects who can address specific health and safety regulations and construction codes and standards. The NIH design manual is a useful reference for the design, construction, commissioning, and maintenance of biomedical laboratories and gives an overview of various design considerations as well as specific information on infrastructure considerations [4].

2.1   Infrastructure Considerations

The build-out of new laboratory space or the renovation of existing laboratory space requires careful planning and programming to achieve the overall goals with design and safety guidelines ensuring that there is sufficient space available to achieve the program goals. The type of experiments (workflow), the number of people, and the weight and size (dimensions) of the equipment provide the detail to inform on the infrastructure requirements and laboratory design. Infrastructure issues are often categorized as either mechanical or electrical.

The mechanical engineering issues address plumbing including water pipes, gas lines, and heating, ventilation, and air conditioning (HVAC). Heating, ventilation, and air conditioning (HVAC) are vitally important to maintaining a healthy and productive work environment. Without properly balanced airflow, the biosafety cabinets may not work optimally, and heat and humidity can build in the laboratory to uncomfortable levels. Furthermore, the balancing of airflow to provide positive or negative air pressures depending on the laboratory function, such as positive pressure in the tissue culture laboratory, can take time and should not be overlooked.

The electrical engineering addresses the placement or upgrade of electrical power including uninterrupted power outlets (emergency power) as well as building management systems, which monitor and control the HVAC, equipment and alarms, and security systems and controlled access.

The lists below identify many of the commonly encountered mechanical and electrical needs for establishing a stem cell laboratory, but they are by no means exhaustive. The architect/engineer team will provide further guidance, and the suggested reading at the end of this chapter provides more information.

2.1.1 Mechanical Engineering

- Plumbing:
  - Sinks with hot and cold water supply:

    Acid resistant

    Faucets with wrist lever and hand-free operation

    Eyewash

    High-purity distilled, reverse osmosis (RO) water or Milli-Q system
  - Safety deluge shower.
  - Ice flake machines and autoclaves, which require easy access to floor drains and water inlet points.
  - Gas line of medical grade copper tubing for CO_2 is an option depending on budget and laboratory design, useful if space permits keeping CO_2 tanks in a separate room (tank farm) rather than inside the tissue culture laboratory.
- Heating, ventilation, and air conditioning (HVAC):
  - Temperature and humidity control
- Fire protection:
  - Sprinkler system, where appropriate

2.1.2 Electrical Engineering

- Lighting
- Lighting—emergency and egress
- Power—outlet type and placement to accommodate equipment
- Power—emergency/uninterruptible
- Fire alarms
- Telephone/data
- Building management systems:

- Building automation systems (BAS)
- Electronic security systems
- Equipment operation and alarm system

2.2 Design Considerations

The cell laboratory design will support the program goals where the type of experiments, number of people, and budget are all considered by the design team. Three basic functional areas are required to optimally run and manage a stem cell laboratory: office or desk space for reading, planning, evaluating data, and reporting; cell culture laboratory where the cells are cultured and maintained; and the molecular biology/quality control laboratory where the quality of the cells is evaluated. In addition to these three functional areas within the stem cell unit or laboratory, access to other support functions is highly desirable, such as a quarantine culture room, adequate storage for supplies, and long-term cell storage in a liquid nitrogen (LN2) freezer room. Finally, the program should have access to shared resources having other specialized equipment and expertise.

2.2.1 Office or Desk Space

The office and desk space required for planning experiments, ordering supplies, reviewing data, and writing reports and papers is an often overlooked but important component of a well-designed laboratory. This space should provide both private, semiprivate, and open collaborative meeting spaces. The amount of administrative space required depends greatly upon the research program. If the program is purely discovery, there may not be as great of a need for administrative space; however, as cellular therapies move toward the clinic, more oversight and more sophisticated quality assurance programs are required and require more paperwork and record-keeping. Again, your architect or facility planner will assist with this, but 10–30% of overall workspace is generally allocated to office and desk space.

2.2.2 Cell Culture Laboratory

The heart of the stem cell laboratory, the cell culture laboratory, generally requires the greatest amount of thought and planning. The large and heavy biosafety cabinets (BSCs) and incubators are not easily moved once in place and require specialized consideration in terms of workflow and backup power supply. Additionally, there are usually specific electric and mechanical requirements associated with the establishment of the cell culture laboratory.

Special air handling requirements to accommodate the air changes and heat generated by the biosafety cabinets and positive pressure to help eliminate the inflow of airborne contaminants are all design considerations. The placement of the BSC, especially when multiple BSCs are to be utilized, is one of the most important design considerations in the cell culture laboratory. The placement of the BSC should allow for a work zone around the BSC, it should not be placed near an entryway, and air supply diffusers or exhaust vents should not be placed directly over or in front of the

BSC. An excellent description of the best practice for placement of BSC can be found in Appendix I of the NIH's Design Requirements Manual [5].

2.2.3 Equipment: Cell Culture Laboratory

- Biosafety cabinet (BSC), [1]Class II, HEPA-filtered exhaust, and downdraft air
- CO_2 incubator
- Microscope—phase/contrast with 4, 10, 20× objectives
- Refrigerator (4 °C)
- Vacuum pump or in-house vacuum system protected with filtration (0.3 micron hydrophobic filter or the equivalent to minimize contamination of vacuum pumps)
- Centrifuge
- Water bath

2.2.4 Molecular Biology/ Quality Control Laboratory

Ideally, the molecular biology/quality control laboratory is located adjacent but outside of the cell culture laboratory and has its own equipment, pipettors, and, perhaps, personnel. With the nature of the assays, and especially cloning activities, that may require the culture of bacteria, the segregation of work and equipment, large and small, is highly recommended. If space or financial limitations do not permit the development of a separate molecular biology laboratory, these the molecular biology and quality control activities should be segregated from the cell culture activities. Section 2.2.5 lists the equipment required for the establishment of the quality control laboratory.

2.2.5 Equipment: Molecular Biology/Quality Control Laboratory

- Chemical fume hood
- Acid and corrosive vented storage cabinet, under the fume hood
- Flammable storage cabinet
- Storage for emergency equipment
- Freezers −20 °C, −80 °C
- Benchtop and floor model centrifuges
- Water bath(s)
- Balance(s)
- Vacuum pump or in-house vacuum system
- PCR thermocycler
- Flask shaker
- LN2 freezers
- Liquid nitrogen supply

---

[1] Three classes of BSC are available, Class II is suitable for Biosafety Level 2 work, and if higher Biosafety Level is required, a Class III BSC may be required.

2.2.6 Laboratory Finishes

Laboratory finishes such as those listed below are recommended by the US General Services Administration (GSA) [6], which is an independent agency of the US government that helps manage and support the basic functioning of government agencies. Other finishes may be used if impervious to common laboratory reagents, able to withstand the cleaning solutions and not prone to harboring contaminants such as bacteria and fungi. The design team will very likely evaluate finishes from the risk, safety, and cost standpoint and make recommendations to the investigator. An example of a substitution material might be the use of plastic laminate for countertops on benchtops, shelving, and other light-duty surfaces. Notably, it must be replaced if delamination occurs, which is a consequence of moisture seepage into the underlying wood-based substrate [7]:

- Wall: epoxy paint, base integral with floor material, steel corner guards
- Floor: monolithic seamless chemical resistant vinyl flooring with integral coved base
- Ceiling: Mylar finish 24″ w by 48″ ceiling tile in suspended ceiling grid
- Casework: pre-manufactured laboratory metal casework system, adjustable shelving with dust cap
- Counter: epoxy resin with integral splash
- Sinks: stainless steel
- Chase system with metal channel support, with horizontal distribution of utilities
- Chairs: vinyl

2.2.7 Shared Support

- Flow Cytometry Core
- Microscopy Core
- Proteomics Core
- DNA Core
- Autoclave
- Glass Wash Room
- Animal facility/vivarium

2.2.8 Storage and Monitoring of Supplies and Reagents

A planned process, part of the quality management system, for receipt, storage, and the staging of materials and supplies should be established. This will allow for the efficient use of the resources. A key objective of close monitoring of supplies and reagents permits the evaluation of and release of production lots of materials such as sera, cytokines, and antibodies that may have variable activities with each lot and permit the evaluation of the performance of biological reagents before they are released into the cell culture laboratory.

2.2.9 Quarantine Laboratory

The introduction of new cell lines or biological reagents can be a major source of contamination in the cell culture laboratory. New cell lines and reagents should be carefully evaluated in a separate quarantine laboratory having a dedicated biosafety cabinet and incubator if possible. If an isolated quarantine laboratory is not available, the new materials can be handled within the tissue culture laboratory by dedicated personnel, hood, and incubator. These unproven materials should be handled at the end of the day, and staff should wear appropriate protective gear, lab coat, gloves, and disposable sleeves, all of which are either disposed of or laundered at the end of each day, i.e., not used next morning for culturing of established cells and reagents.

If the cultures come to the laboratory in a medium containing antibiotics, at the time of the first subculture, passage some of the cells into growth medium that does not contain antibiotics while maintaining the rest of the culture in antibiotic-containing medium. Cells grown in the antibiotic-free medium should then be tested for bacterial, fungal, and mycoplasma contamination.

3 Operations Overview

Facilities and proper equipment are necessary, but not sufficient, for the operation of the stem cell laboratory. Skilled personnel and established protocols and/or standard operating procedures (SOPs) help ensure efficient and effective operations of routine activities.

Laboratory operations are greatly improved through the development of quality assurance plan wherein the operations are planned, reviewed, and then performed. This approach is highly effective for routine techniques and procedures and supports high-quality, reproducible research activities. The extent to which a quality system is developed and implemented depends on the stage of the research; as the research advances to clinical applications, more attention to quality assurance is needed, if not required.

3.1 Production and Process Controls

A state-of-the-art stem cell laboratory will have in place a quality assurance system that permits the oversight of cell line production with established protocols or SOPs relating to the receipt or derivation, culture and maintenance, and banking and storage of cell lines as well as the associated quality control assays. Whether the research program is aimed at the development of cellular therapies or basic discovery, the incorporation of a quality systems approach will greatly facilitate laboratory operations and provide a solid foundation for future experiments as well as set the stage for quality preclinical and clinical studies [8–10].

3.1.1   Cells

Careful documentation of the source of the cell line, cell type, tissue source, passage number, growth medium, growth conditions, and subculture methodology should be performed for each line brought into the laboratory. The provenance of the cell lines should be carefully documented.

Culturing and preparing cell lines for experimentation while maintaining cell line stability and keeping them free from cross contamination by other cell lines and contamination by bacterial, fungal, or mycoplasma is one of the primary goals of the stem cell laboratory. Toward that end, the development of protocols and SOPs that permit the consistent and reliable culture of each cell line greatly facilitates laboratory operations.

3.1.2   Reference Standards

One of the major challenges in establishing a new stem cell laboratory is often the acquisition of well-characterized stem cell lines that can serve as the reference stocks for testing new equipment, training staff, and developing new techniques. Obtaining a qualified standard cell line to serve as a laboratory reference standard will provide a solid foundation for establishing procedures and assays [11, 12]. The use of reference materials can accelerate the establishment of the laboratory and generate internal and external confidence in the program.

3.1.3   Cell Banking

The establishment of a stem cell bank with known passage numbers and/or population doublings can be critical to the overall success of the stem-based program. Most stem cell lines have a defined range of passages wherein they are most stable and most useful for experimentation. Stem cells may gain mutations and lose primary characteristics such as potency or differentiation capacity with prolonged culture [13]. It is critical to the development of a successful banking system to know the stability of the cell type under the planned culture conditions. A tiered banking system that utilizes master cell bank from which working cell banks are produced allows for maximum expansion and long-term experimental reproducibility.

The tiered banking system provides constant supply of the starting material and detailed characterization of the cell line decreasing the likelihood and increasing the detection of cell line cross contamination and adventitious agent contamination [8].

The establishment of a proper cell banking and documentation system establishes the foundation upon which experimentation can take place ensuring long-term success of the stem cell program. At any time, the researcher can confidently return to proven cell stocks for use in experimentation and confirmation.

Each vial of the master cell bank (MCB) can be used to generate the working cell bank (WCB), from which each vial can be used for a determined number of passages, based on the stability profile.

Cell Seed Lot System

- Seed stock/research bank <10 vials
- Master cell bank (MCB) 20–200 vials
- Working cell bank (WCB) 50–500 vials

3.1.4 Reagents and Materials

Most of the reagents and materials used in the cell laboratory can be obtained from commercial vendors with established manufacturing processes with certificates of assurance (CoA) that describe the testing that has been performed to ensure sterility and, perhaps, activity. For those reagents that do not come from commercial sources or have lot-to-lot variability, such as sera, antibodies, and some cytokines, it will be necessary to establish qualifying assays with acceptable ranges of activities. These reagents, usually biologically sourced, may require quarantine prior to release into the laboratory for use.

3.1.5 Quality Control Assays

The use of quality control assays at key points during routine culture, the time of cell banking, and as part of the routine laboratory operations is key to the production of high-quality data and reproducible experiments that can form the foundation of preclinical studies and clinical translation of the cell product, if that is part of the program goal.

The frequency and timing of the initiation of the quality control assay panel should be determined based on an event such as cell banking or the identification of a contaminated culture or as part of the routine operations of the laboratory, such as every ten passages or every 3 months.

The exact assays which comprise the quality control panel are cell-type specific, but fall into four major categories of sterility, identity, purity, and potency [9]. A general overview of the considerations for each is provided below.

Sterility/Microbiological Testing

Maintaining a sterile culture is the foundation upon which other activities of the cell laboratory are based. Frequent observation of cultures for changes in medium color, medium turbidity, cell morphology, and growth rate can be clues that something has gone awry in the culture. Trying to determine the root cause of the contamination can be a useful process where additional training can be implemented or a contaminated reagent can be eliminated. It is not advisable to try to "cure" the culture. Unless it is a very valuable, unreplaceable cell line, it is best to dispose of this culture; thoroughly clean the hood, incubator, microscope stage, etc.; and initiate a new culture from the established working cell bank or go all the way back to the cell source.

Fungal and Bacterial Sterility Testing

This is usually done by classical culture techniques that use fluid thioglycollate medium (Sigma catalog no. 90404) for detection of aerobic and anaerobic bacteria and soya bean/casein broth (Sigma catalog no. S1674) for detection of aerobes, facultative anaerobes, and fungi. Samples are inoculated into the broths in duplicate, with one being cultured at 22 °C and the other at 32 °C for 14 days, after which they are examined for turbidity.

Mycoplasma Testing

Mycoplasma is a common contaminant in mammalian cell culture and can be very insidious. Cultures can be contaminated with mycoplasma from the reagents or from the technician, and once contaminated, low levels may persist undetected. Frequent testing, monthly, and quarantine of new cell lines can help prevent mycoplasma, should it be found, from spreading.

Adventitious Viral Agent Testing

Screening a new cell line or tissue of human origin for the following human viruses may be necessary: CMV, HIV-1, HIV-2, EBV, B19, HBV, and HCV. Depending on the programmatic goals, screening cell banks for animal viruses may be necessary as well. Bovine spongiform encephalopathy (BSE), the virus that causes "mad cow disease," is a concern for cells that have been cultivated in bovine serum-containing medium. Unfortunately, there is no screen for BSE—a judicious choice of suppliers keeps this risk to a minimum.

Identity

Maintaining pure cultures free from contamination by other cell lines is a primary function of the cell culture laboratory and of utmost importance when performing experiments using this stem cell line. Confirming the identity of the culture and distinguishing it from other cells being cultured in the laboratory are goals of the assays that determine cell line identity [9].

Short tandem repeat (STR) genetic analysis can be used to authenticate human cell lines [14]. Short tandem repeats (STRs) are polymorphic regions of genomic DNA that are used for human identity testing. A core set of STRs has been developed, and the primer sequences are available online through the National Institute of Standards and Testing (NIST) website, STRBase (www.cstl.nist.gov/strbase).

Cytogenetic techniques used to determine the number of chromosomes and chromosomal banding patterns can help identify different species, such as the mouse, monkey, and human.

Purity

In the case of stem cell cultures, purity generally refers to the homogeneity of the culture [9]. The assays used to determine purity should determine whether the culture is free from unwanted extraneous cell types. Purity is often determined using flow cytometry, immunohistochemistry, PCR, or multiplex PCR to determine the number of cells expressing the desired cell-specific markers in relation to those cells either not expressing those markers or markers indicating contaminating cell population(s) within the culture [15, 9]. Cytogenetic analysis of the culture can also be performed to determine the karyotype of the culture [16, 10].

Potency

The development of a small panel of assays that can be used to assess the potency of the cells is important as well. The potency assays are usually functional assays that include in vitro and in vivo analysis of function [9]. In vitro assays may include expression of cytokines or in vitro differentiation. In vivo function may include functional analysis in an animal model of disease [17, 18] or assessment of potential such as general differentiation potential based on teratoma formation [19–21].

Stability

The stability of stem cell cultures is a major concern with many reports showing changes in cells over time in culture [22, 23]. Some changes such as epigenetic or small deletions and insertions are subtle and cannot be easily determined without costly assays and expensive equipment [13, 24]. Others such as aneuploidy can be identified through routine quality control assays which evaluate the karyotype of the cell line. Careful observation of the cultures for changes in morphology, growth rate, differentiation performance during routine cultivation, and experimentation is often the first indication that the cells may not be as expected.

The optimal range of passages or population doublings should be determined experimentally for each stem cell type in use in the program and the cells should be banked at passages that account for this limitation.

3.1.6   Contamination Control

It is extremely difficult to eliminate microbial contamination from cell cultures; it is best to put in place controls that help to prevent contamination from occurring and remain highly vigilant for any changes in the cultures that could signal contamination. These controls are in the form of properly trained staff aseptic technique and proper use of the biosafety cabinet, cleaning and maintenance of equipment and the laboratory, and quarantine of newly arrived materials, especially cells and tissues.

References

1. Wesselschmidt RL, Schwartz PH (2011) The stem cell laboratory: design, equipment, and oversight. Methods Mol Biol 767:3–13. doi:10.1007/978-1-61779-201-4_1

2. Lyons I, Tan D, Schwartz PH, Rao M (2007) Setting up a laboratory. In: Loring JF, Wesselschmidt RL, Schwartz PH (eds) Human stem cell manual: a laboratory guide, 1st edn. Elsevier, New York

3. Freshney RI (2005) Laboratory design and layout. In: Freshney RI (ed) Culture of animal cells: a manual of basic technique, 5th edn. John Wiley & Sons, Inc., Hoboken, New Jersey, pp 43–53

4. Yamanaka K, Chun SJ, Boillee S, Fujimori-Tonou N, Yamashita H, Gutmann DH, Takahashi R, Misawa H, Cleveland DW (2008) Astrocytes as determinants of disease progression in inherited amyotrophic lateral sclerosis. Nat Neurosci 11(3):251–253. doi:10.1038/nn2047

5. Farhad Memarzadeh (2010) Biosafety Cabinet (BSC) placement requirements for new buildings and renovations. In: Appendix I. NIH

6. Bensimon G, Lacomblez L, Meininger V (1994) A controlled trial of riluzole in amyotrophic lateral sclerosis. ALS/Riluzole Study Group. N Engl J Med 330(9):585–591. doi:10.1056/NEJM199403033300901

7. Laboratory Plastic Laminate (2015). News to use, Design Requirements Manual 01:1

8. Guidance for Industry: Guidance for Human Somatic Cell Therapy and Gene Therapy (1998). US Department of Health and Human Services, FDA, CBER

9. Guidance for Industry Content and Review of Chemistry, Manufacturing, and Control (CMC), Information for Human Somatic Cell Therapy Investigational New Drug Applications (INDs) (2008). US Department of Health and Human Services, FDA, CBER

10. Andrews PW, Baker D, Benvinisty N, Miranda B, Bruce K, Brüstle O, Choi M, Choi YM, Crook JM, de Sousa PA, Dvorak P, Freund C, Firpo M, Furue MK, Gokhale P, Ha HY, Han E, Haupt S, Healy L, Hei DJ, Hovatta O, Hunt C, Hwang SM, Inamdar MS, Isasi RM, Jaconi M, Jekerle V, Kamthorn P, Kibbey MC, Knezevic I, Knowles BB, Koo SK, Laabi Y, Leopoldo L, Liu P, Lomax GP, Loring JF, Ludwig TE, Montgomery K, Mummery C, Nagy A, Nakamura Y, Nakatsuji N, Oh S, Oh SK, Otonkoski T, Pera M, Peschanski M, Pranke P, Rajala KM, Rao M, Ruttachuk R, Reubinoff B, Ricco L, Rooke H, Sipp D, Stacey GN, Suemori H, Takahashi TA, Takada K, Talib S, Tannenbaum S, Yuan BZ, Zeng F, Zhou Q (2015) Points to consider in the development of seed stocks of pluripotent stem cells for clinical applications: International Stem Cell Banking Initiative (ISCBI). Regen Med 10(2s):1–44. doi:10.2217/rme.14.93

11. French A, Bravery C, Smith J, Chandra A, Archibald P, Gold JD, Artzi N, Kim H-W, Barker RW, Meissner A, Wu JC, Knowles JC, Williams D, García-Cardeña G, Sipp D, Oh S, Loring JF, Rao MS, Reeve B, Wall I, Carr AJ, Bure K, Stacey G, Karp JM, Snyder EY, Brindley DA (2015) Enabling consistency in pluripotent stem cell-derived products for research and development and clinical applications through material standards. Stem Cells Transl Med 4(3):217–223. doi:10.5966/sctm.2014-0233

12. Tanavde V, Vaz C, Rao MS, Vemuri MC, Pochampally RR (2015) Research using mesenchymal stem/stromal cells: quality metric towards developing a reference material. Cytotherapy 17(9):1169–1177. doi:10.1016/j.jcyt.2015.07.008

13. Laurent LC, Ulitsky I, Slavin I, Tran H, Schork A, Morey R, Lynch C, Harness JV, Lee S, Barrero MJ, Ku S, Martynova M, Semechkin R, Galat V, Gottesfeld J, Belmonte JCI, Murry C, Keirstead HS, Park H-S, Schmidt U, Laslett AL, Muller F-J, Nievergelt CM, Shamir R, Loring JF (2011) Dynamic changes in the copy number of pluripotency and cell proliferation genes in human ESCs and iPSCs during reprogramming and time in culture. Cell Stem Cell 8(1):106–118. doi:10.1016/j.stem.2010.12.003

14. Barallon R, Bauer SR, Butler J, Capes-Davis A, Dirks WG, Elmore E, Furtado M, Kline MC, Kohara A, Los GV, MacLeod RAF, Masters JRW, Nardone M, Nardone RM, Nims RW, Price PJ, Reid YA, Shewale J, Sykes G, Steuer AF, Storts DR, Thomson J, Taraporewala Z, Alston-Roberts C, Kerrigan L (2010) Recommendation of short tandem repeat profiling for authenticating human cell lines, stem cells, and tissues. In Vitro Cell Dev Biol Anim 46(9):727–732. doi:10.1007/s11626-010-9333-z

15. Mamidi MK, Pal R, Bhonde R, Zakaria Z, Totey S (2010) Application of multiplex PCR for characterization of Human Embryonic Stem Cells (hESCs) and its differentiated progenies. J Biomol Screen 15(6):630–643. doi:10.1177/1087057110370211

16. Wesselschmidt RL, Loring JF (2007) Classical cytogenetics: karyotyping. In: Loring JF, Wesselschmidt RL, Schwartz PH (eds) Human stem cell manual: a laboratory guide. Elsevier, New York, NY, pp 59–70

17. Mendez DC, Stover AE, Rangel AD, Brick DJ, Nethercott HE, Torres MA, Khalid O, Wong AMS, Cooper JD, Jester JV, Monuki ES, McGuire C, Le SQ, S-h K, Dickson PI, Schwartz PH (2015) A novel, long-lived, and highly engraftable immunodeficient mouse model of mucopolysaccharidosis type I. Mol Ther Methods Clin Dev 2:14068. doi:10.1038/mtm.2014.68

18. Priest CA, Manley NC, Denham J, Wirth ED, Lebkowski JS (2015) Preclinical safety of human embryonic stem cell-derived oligodendrocyte progenitors supporting clinical trials in spinal cord injury. Regen Med 10(8):939–958. doi:10.2217/rme.15.57

19. Wesselschmidt R (2011) The teratoma assay: an in vivo assessment of pluripotency. In: Schwartz PH, Wesselschmidt RL (eds) Human pluripotent stem cells, Methods in molecular

biology, vol 767. Humana Press, pp 231–241. doi:10.1007/978-1-61779-201-4_17

20. Damjanov I, Ahrlund-Richter L (2007) Characterization of human embryonic stem cell derived teratomas. In: Loring JF, Wesselschmidt RL, Schwartz PH (eds) Human stem cell manual: a laboratory guide, 1st edn. Elsevier, New York

21. Gertow K, Przyborski S, Loring JF, Auerbach JM, Epifano O, Otonkoski T, Damjanov I, Ährlund-Richter L (2007) Isolation of human embryonic stem cell–derived teratomas for the assessment of pluripotency. In: Current protocols in stem cell biology. John Wiley & Sons, Inc. doi:10.1002/9780470151808.sc01b04s3

22. Xie X, Hiona A, Lee AS, Cao F, Huang M, Li Z, Cherry A, Pei X, Wu JC (2011) Effects of long-term culture on human embryonic stem cell aging. Stem Cells Dev 20(1):127–138. doi:10.1089/scd.2009.0475

23. Pan X, Li X-J, Liu X-J, Yuan H, Li J-F, Duan Y-L, Ye H-Q, Fu Y-R, Qiao G-H, Wu C-C, Yang B, Tian X-H, Hu K-H, Miao L-F, Chen X-L, Zheng J, Rayner S, Schwartz PH, Britt WJ, Xu J, Luo M-H (2013) Later passages of neural progenitor cells from neonatal brain are more permissive for human cytomegalovirus infection. J Virol 87(20):10968–10979. doi:10.1128/JVI.01120-13

24. Garitaonandia I, Amir H, Boscolo FS, Wambua GK, Schultheisz HL, Sabatini K, Morey R, Waltz S, Wang Y-C, Tran H, Leonardo TR, Nazor K, Slavin I, Lynch C, Li Y, Coleman R, Gallego Romero I, Altun G, Reynolds D, Dalton S, Parast M, Loring JF, Laurent LC (2015) Increased risk of genetic and epigenetic instability in human embryonic stem cells associated with specific culture conditions. PLoS One 10(2):e0118307. doi:10.1371/journal.pone.0118307

INDEX

Amit K. Srivastava et al. (eds.), *Stem Cell Technologies in Neuroscience*, Neuromethods, vol. 126,
DOI 10.1007/978-1-4939-7024-7, © Springer Science+Business Media LLC 2017

Printed in the United States
By Bookmasters